AF352840

Recent Trends in Computational Research on Diseases

Recent Trends in Computational Research on Diseases

Editors

Md. Altaf-Ul-Amin
Shigehiko Kanaya
Naoaki Ono
Ming Huang

MDPI • Basel • Beijing • Wuhan • Barcelona • Belgrade • Manchester • Tokyo • Cluj • Tianjin

Editors
Md. Altaf-Ul-Amin
Nara Institute of Science and
Technology
Japan

Shigehiko Kanaya
Nara Institute of Science and
Technology
Japan

Naoaki Ono
Nara Institute of Science and
Technology
Japan

Ming Huang
Nara Institute of Science and
Technology
Japan

Editorial Office
MDPI
St. Alban-Anlage 66
4052 Basel, Switzerland

This is a reprint of articles from the Special Issue published online in the open access journal *Life* (ISSN 2075-1729) (available at: https://www.mdpi.com/journal/life/special_issues/computational_diseases).

For citation purposes, cite each article independently as indicated on the article page online and as indicated below:

LastName, A.A.; LastName, B.B.; LastName, C.C. Article Title. *Journal Name* **Year**, *Volume Number*, Page Range.

ISBN 978-3-0365-3230-1 (Hbk)
ISBN 978-3-0365-3231-8 (PDF)

Contents

About the Editors

Md. Altaf-Ul-Amin received his B.Sc. degree in Electrical and Electronic Engineering from Bangladesh University of Engineering and Technology (BUET), Dhaka, his M.Sc. degree in Electrical, Electronic and Systems Engineering from University Keban gsaan Malaysia (UKM) and his PhD degree from the Nara Institute of Science and Technology (NAIST), Japan. He previously worked in several universities in Bangladesh, Malaysia, and Japan. Currently he is working as an associate professor in the Computational Systems Biology Lab of NAIST. His major research topics are Network Biology, Systems Biology, Cheminformatics and Biological Databases. He developed and implemented novel graph clustering algorithms DPClus and DPClusO.

Shigehiko Kanaya received his B.Sc. degree in Bioscience from the Science University of Tokyo, Japan in 1985, and his Ph.D. from Toyohashi University of Technology, Japan in 1990. He served as an Assistant Professor in Information Engineering at Yamagata Univ. in 1990, Guest Associate Professor at the National Institute of Genetics in 1996, Associate Professor at Electronic and Information Engineering in 1999, Associate Professor, Applied Bio System Engineering at Yamagata Univ. in 2000, Guest researcher at Bio radical Institute (Yamagata Prefecture), in 2000 Associate Professor, Research and Education Centre for Genetic Information at NAIST in 2001, Associate Professor, Graduate School of Information Science at NAIST in 2002, Professor, Graduate School of Information Science at NAIST since 2004. Developed KNApSAcK family databases including a comprehensive species-metabolite relational database. His research areas are systems biology, biological databases, metabolomics.

Naoaki Ono Completed his PhD at the University of Tokyo in 2001. After working as a CREST researcher at Kyoto University in 2001, he moved to the Advanced Telecommunications Research Institute International. Since 2007, he has served as an ERATO researcher at Osaka University and since 2012, he has been an assistant professor at the Graduate School of Information Science, Nara Institute of Science and Technology (NAIST). Since 2018, he has been an associate professor at the Data Science Centre in NAIST. His specialty is theanalysis and simulation of life phenomena and their evolution based on bioinformatics, systems biology, complex system models, using statistical models and deep learning models.

Ming Huang received Ph.D. degree in biomedical engineering in 2012 from the University of Aizu, Aizu, Japan. He is currently an assistant professor at the Nara Institute of Science and Technology and a visiting scholar at the Biomedical Engineering department of the University of California Davis. His research interests include AI engineering for digital health management systems, and digital biomarker identification, which are indispensable parts of health informatics. He is also interested in applying graph theory and the relevant graph neural network in quantitative structure-activity relationship (QSAR) studies for drug discovery.

Editorial

Recent Trends in Computational Biomedical Research

Md. Altaf-Ul-Amin *, Shigehiko Kanaya, Naoaki Ono and Ming Huang

Division of Information Science, Nara Institute of Science and Technology, Ikoma 630-0192, Japan;
skanaya@gtc.naist.jp (S.K.); nono@is.naist.jp (N.O.); alex-mhuang@is.naist.jp (M.H.)
* Correspondence: amin-m@is.naist.jp

Citation: Altaf-Ul-Amin, M.; Kanaya, Ono, N.; Huang, M. Recent Trends Computational Biomedical Research. *Life* **2022**, *12*, 27. https:// .org/10.3390/life12010027

Received: 23 December 2021
Accepted: 23 December 2021
Published: 24 December 2021

Publisher's Note: MDPI stays neutral with regard to jurisdictional claims in published maps and institutional affiliations.

Recent advances in information technology have brought forth a paradigm shift in science, especially in the biology and medical fields. Statistical methodologies based on high-performance computing and big data analysis are now indispensable for qualitative and quantitative understanding of experimental results. In fact, the last few decades have witnessed drastic improvements in high-throughput experiments in health science, for example, mass spectrometry, DNA microarray, and next-generation sequencing. Those methods have been providing massive data involving four major branches of omics (genomics, transcriptomics, proteomics, and metabolomics). On the other hand, cell imaging, clinical imaging, and personal healthcare devices are also providing important data concerning the human body and disease. In parallel, various methods of mathematical modeling such as machine learning have also developed rapidly. All of the types of these data can be utilized in computational approaches for biomedical research such as on understanding disease mechanisms, diagnosis, prognosis, drug discovery, drug repositioning, disease biomarkers, driver mutations, copy number variations, disease pathways, and much more. Therefore, the range of topics under "Recent Trends in Computational Biomedical Research" is extensive, and the present Special Issue is not a comprehensive representation of the subject. Nonetheless, the articles selected for this Special Issue represent a variety of topics relating to the title, and we are sharing with the readers with pleasure.

In this Special Issue, we published a total of eight good papers. Overall, four papers are in cardiovascular-related topics, two are linked to drug development, and the other two are on finding associations between genome sequence aberrations and diseases.

The paper titled "SMCKAT, a Sequential Multi-Dimensional CNV Kernel-Based Association Test" is on copy number variants (CNVs) [1]. Associations between CNVs and various diseases have been well studied before, and the current paper proposes a method called SMCKAT to test the association between the sequential order of CNVs and diseases. SMCKAT was evaluated by applying on CNV data sets, demonstrating its ability to exhibit rare or common CNVs by detecting specific biologically relevant chromosomal regions supported by the biomedical literature.

The paper with the title "The Genome-Wide Scanning of Potential Hotspots for Adenosine Methylation: A Potential Path to Neuronal Development" is about the methylation of adenosines at the N6 position (m6A) [2]. This methylation is the most frequent type of internal modification in mRNAs and is attributable to diverse roles in physiological developments and pathophysiological processes. This work applied a sliding window technique to identify the consensus site and annotated all m6A hotspots in the human genome. Most of the hotspots were found to be located in the non-coding regions, suggesting that methylation occurs before splicing. The role of this methylation in neuron physiology was also elaborated

An automatic ECG quality assessment method has been presented in the paper titled "Electrocardiogram Quality Assessment with a Generalized Deep Learning Model Assisted by Conditional Generative Adversarial Networks" [3]. This automatic method can help cardiologists perform diagnosis much faster. The proposed system first trained a conditional generative adversarial network model for data augmentation and then developed a deep quality assessment model based on a training dataset composed of real and generated ECG. The proposed system demonstrated improved performance in the ECG quality assessment and has the potential to be utilized in clinical practice.

The paper titled "A Maximum Flow-Based Approach to Prioritize Drugs for Drug Repurposing of Chronic Diseases" is on drug repurposing, i.e., using existing drugs to treat new/other diseases [4]. The work proposes a maximum-flow-based approach using protein-protein interactions (PPIs) of drug targets (proteins) and risk genes corresponding to chronic diseases such as breast cancer, inflammatory bowel disease (IBD), and chronic obstructive pulmonary disease (COPD). The top candidate drugs identified by the maximum flow-based approach were found to be experimentally validated by other independent studies.

Models and results of in silico prediction of the interactions between compounds of Jamu herbs and human proteins have been presented in the paper titled "Identification of Targeted Proteins by Jamu Formulas for Different Efficacies Using Machine Learning Approach" [5]. Verifying the proteins that are targeted by natural compounds is helpful to select natural herb-based drug candidates and to explain the scientific mechanisms of traditional medicines.

The paper titled "Shared Molecular Mechanisms of Hypertrophic Cardiomyopathy and Its Clinical Presentations: Automated Molecular Mechanisms Extraction Approach is about finding molecular mechanisms of hypertrophic cardiomyopathy (HCM), which is the most common inherited cardiovascular disease [6]. Molecular mechanisms were extracted from abstracts and open access full articles by multiple machine-reading systems. Shared molecular mechanisms of HCM and its clinical presentations were represented as networks, and the most important elements in the networks were found based on node centrality measures.

The paper with the title "Impact of Water Temperature on Heart Rate Variability during Bathing" aims to explore the impact of water temperature (WT) on Heart Rate Variability (HRV) by collecting five electrocardiogram (ECG) recordings of each of 10 subjects at different preset bathtub WT conditions during bathing [7]. Based on statistical analysis, it has been shown that the WT has an important impact on HRV during bathing.

The accuracy of several methods has been compared and assessed in the paper with the title "Discussion of Cuffless blood pressure prediction using plethysmograph based on a longitudinal experiment: Is individual model necessary?" [8]. Estimating blood pressure using the Plethysmograph (PPG) signal is convenient and makes continuous measurement feasible, but some doubt exists on its accuracy and robustness. By comparing the regression models, this paper shows that an individual Gaussian Process model attains the best result.

Funding: This research received no external funding.

Acknowledgments: Our heartfelt thanks are due to authors for their excellent and thought-provoking contributions and their patience in communicating with us. Finally, we acknowledge all the dedicated reviewers of these papers for their critical and insightful comments.

Conflicts of Interest: The authors declare no conflict of interest.

References

1. Maus Esfahani, N.; Catchpoole, D.; Kennedy, P.J. SMCKAT, a Sequential Multi-Dimensional CNV Kernel-Based Association Test. *Life* **2021**, *11*, 1302. [CrossRef]
2. Kumar, S.; Tsai, L.-W.; Kumar, P.; Dubey, R.; Gupta, D.; Singh, A.K.; Swarup, V.; Singh, H.N. Genome-Wide Scanning of Potential Hotspots for Adenosine Methylation: A Potential Path to Neuronal Development. *Life* **2021**, *11*, 1185. [CrossRef] [PubMed]
3. Zhou, X.; Zhu, X.; Nakamura, K.; Noro, M. Electrocardiogram Quality Assessment with a Generalized Deep Learning Model Assisted by Conditional Generative Adversarial Networks. *Life* **2021**, *11*, 1013. [CrossRef] [PubMed]
4. Islam, M.M.; Wang, Y.; Hu, P. A Maximum Flow-Based Approach to Prioritize Drugs for Drug Repurposing of Chronic Diseases. *Life* **2021**, *11*, 1115. [CrossRef] [PubMed]
5. Wijaya, S.H.; Afendi, F.M.; Batubara, I.; Huang, M.; Ono, N.; Kanaya, S.; Altaf-Ul-Amin, M. Identification of Targeted Proteins by Jamu Formulas for Different Efficacies Using Machine Learning Approach. *Life* **2021**, *11*, 866. [CrossRef] [PubMed]
6. Glavaški, M.; Velicki, L. Shared Molecular Mechanisms of Hypertrophic Cardiomyopathy and Its Clinical Presentations: Automated Molecular Mechanisms Extraction Approach. *Life* **2021**, *11*, 785. [CrossRef] [PubMed]

Xu, J.; Chen, W. Impact of Water Temperature on Heart Rate Variability during Bathing. *Life* **2021**, *11*, 378. [CrossRef] [PubMed]

Kido, K.; Chen, Z.; Huang, M.; Tamura, T.; Chen, W.; Ono, N.; Takeuchi, M.; Altaf-Ul-Amin, M.; Kanaya, S. Discussion of Cuffless Blood Pressure Prediction Using Plethysmograph Based on a Longitudinal Experiment: Is the Individual Model Necessary? *Life* **2022**, *12*, 11. [CrossRef]

Article

Discussion of Cuffless Blood Pressure Prediction Using Plethysmograph Based on a Longitudinal Experiment: Is the Individual Model Necessary?

Koshiro Kido [1], Zheng Chen [1], Ming Huang [1,*], Toshiyo Tamura [2], Wei Chen [3], Naoaki Ono [1,4], Masachika Takeuchi [5], Md. Altaf-Ul-Amin [1] and Shigehiko Kanaya [1,4]

[1] Graduate School of Science and Technology, Nara Institute of Science and Technology, Ikoma 630-0192, Japan; kido.koshiro.kb3@is.naist.jp (K.K.); chen.zheng.bn1@is.naist.jp (Z.C.); nono@is.naist.jp (N.O.); amin-m@is.naist.jp (M.A.-U.-A.); skanaya@gtc.naist.jp (S.K.)

[2] Institute for Healthcare Robotics, Waseda University, Tokyo 162-0041, Japan; t.tamura3@kurenai.waseda.jp

[3] Department of Electronic Engineering, School of Information Science and Technology, Fudan University, Shanghai 201203, China; w_chen@fudan.edu.cn

[4] Data Science Center, Nara Institute of Science and Technology, Ikoma 630-0192, Japan

[5] San-Ei Medisys Co., Ltd., Kyoto 607-8116, Japan; takeuchi@san-ei.com

* Correspondence: alex-mhuang@is.naist.jp; Tel.: +81-743-72-5321

Abstract: Using the Plethysmograph (PPG) signal to estimate blood pressure (BP) is attractive given the convenience and possibility of continuous measurement. However, due to the personal differences and the insufficiency of data, the dilemma between the accuracy for a small dataset and the robustness as a general method remains. To this end, we scrutinized the whole pipeline from the feature selection to regression model construction based on a one-month experiment with 11 subjects. By constructing the explanatory features consisting of five general PPG waveform features that do not require the identification of dicrotic notch and diastolic peak and the heart rate, three regression models, which are partial least square, local weighted partial least square, and Gaussian Process model, were built to reflect the underlying assumption about the nature of the fitting problem. By comparing the regression models, it can be confirmed that an individual Gaussian Process model attains the best results with 5.1 mmHg and 4.6 mmHg mean absolute error for SBP and DBP and 6.2 mmHg and 5.4 mmHg standard deviation for SBP and DBP. Moreover, the results of the individual models are significantly better than the generalized model built with the data of all subjects.

Keywords: blood pressure; cuffless measurement; longitudinal experiment; plethysmograph; nonlinear regression

Citation: Kido, K.; Chen, Z.; Huang, L.; Tamura, T.; Chen, W.; Ono, N.; keuchi, M.; Altaf-Ul-Amin, M.; anaya, S. Discussion of Cuffless ood Pressure Prediction Using ethysmograph Based on a ngitudinal Experiment: Is the dividual Model Necessary? *Life* **22**, *12*, 11. https://doi.org/ 3390/life12010011

ademic Editor: Jonathan L. S. guerra

ceived: 17 November 2021
cepted: 18 December 2021
blished: 22 December 2021

blisher's Note: MDPI stays neutral h regard to jurisdictional claims in lished maps and institutional affil-ons.

1. Introduction

Ambulatory blood pressure (BP) monitoring provides abundant cardiovascular information, and we have seen numerous studies focusing on replacing the conventional auscultatory/oscillometric measurement that requires the occlusion of arterial blood flow cuffless by using the cardiovascular biosignals with state-of-the-art machine learning methods.

The cuffless methods can be roughly categorized into three groups. The first one is based on the pulse arrival time (PAT) [1,2], the second one is based on photoplethysmograph (PPG) signal, which has attracted more and more attention in recent years [3–6], and the third one is based on other methods [7,8].

The PAT is the period that includes the pulse transition time and the pre-ejection period of the heart. Because the PAT needs the ECG signal and distal PPG signal to be determined, it is a modality that is suitable for intermittent measurement.

The second group utilizes only one biosignal—the PPG, which is a signal that reflects the change of blood volume. The pulsatile component of PPG reflects the change in blood volume and the stable component is related to the basic blood volume and other physiological indices such as respiration and body temperature [9]. Since the PPG signal depends on the blood volume in its optical path, which typically covers the arterial and venous capillaries [10], it can be related to the cardiovascular indices such as the blood oxygen saturation and arterial compliance [11,12]. It is generally accepted that the PPG waveform comprises the arterial pulse wave from the left ventricle to the distal sites and the reflected wave from the sites of impedance mismatch [13]. With these physiological understandings, research interest has been shifting to BP estimation based on PPG signal recently. Xing et al. carried out a mass experiment (1249 subjects, 2358 measurements), and by using the PPG waveform features and biometrics in a bagged regression tree model, they got a 9.5 mmHg, 2.2 mmHg, and 17.4 mmHg mean absolute error (MAE) for hypotensive, normotensive, and hypertensive subjects [3]. Chowdhury et al. used an open dataset obtained in a hospital with 219 subjects and 657 measurements to extract the PPG waveform features and biometrics, which were then inputted into the Gaussian Process Regression model. By using a further Bayesian optimization, they claimed 3.02 mmHg and 1.74 mmHg MAE for SBP and DBP estimation, respectively [6].

There are two main criteria for the cuffless method. The first one is that the method/system should meet the requirement on accuracy determined by IEEE Std 1708-2014. In the mean time, when used as a healthcare device, the convenience for personal use should also be taken into account. Therefore, the final goal of development should be an appropriate computation model with high accuracy and long-term stability.

These preliminary studies show the prospect of cuffless BP estimation with PPG signal, however, fundamental issues remain. The first one is the dilemma between the generalization of the regression model and the individual difference of the limited data. This kind of difference is pervasive in biomedical engineering and the method of making a balance is reflected in the modeling assumption and the choice of machine learning model. Data-driven approach resorts to the accumulation of samples to find the relation between explanatory and dependent variables. A globally nonlinear regression model that severely twists the fitting hyperplane to fit the data on hand may not be well applicable for an individual outside the dataset, especially when the size of the dataset is small. It is not uncommon to see the difficulty in getting a reproducible relation between the waveform features and the blood pressure learned from a small dataset in a large population test. The relation may be altered substantially by the demographic factors and physiological/pathological factors. For example, Allen has confirmed the PPG contour triangulation with aging [14]. This means that a robust method should not rely on the identification of dicrotic notch and/or diastolic peak in the PPG waveform.

Moreover, the cardiovascular functions including BP are regularized by biorhythm modulated by the baroreflex and the autonomic nervous system [15,16], which implies the necessity of re-calibration for long-term use.

Based on the discussion above, a through discussion is necessary before the large scale validation experiment. In this study, we tried to contribute to this topic by answering the following two fundamental questions: (*Q1*) When compared with the individual model, the generalized model good enough? (*Q2*) Is it necessary to calibrate in a relatively long term scenario? Special attention is also paid to the following aspects of lift performance (corresponding to *A1*) and robustness (corresponding to *A2*) of the model: (*A1*) to decide the best model for BP regression by comparing the linear, local linear, and Bayesian models; (*A2*) to examine the feasibility of using the general waveform features solely in the regression models, which means that the identification of the dicrotic notch and diastolic peak unnecessary to address the deterioration of the dicrotic notch and the diastolic peak with aging and hypertension.

2. Materials and Methods

To the best knowledge of the authors, there are very few open datasets dedicated the study of cuffless BP estimation. For example, the MIMIC database is well used in blood pressure estimation, whose protocol is not optimized for the BP study. The Non-invasive

Blood Pressure Estimation dataset [17], collected data right after running, which would change the behavior of the baroreflex [18].

In view of the situation that none of the open dataset can be used to answer the questions Q1 and Q2, a one-month experiment, from mid-May to mid-June, dedicated to the development and validation of cuffless BP measurement was conducted. The experiment used a PPG-sensor with 940 nm LED (infrared) of a medical device (Checkme Pro B, San-ei Medisys) for 30 s PPG measurement and a clinical Blood Pressure Monitor (A&D, Model UA-1200BLE) as the reference [19]. Eleven subjects, whose basic information can be found in Table 1, participated in the experiment, during which they were required to measure the PPG signal and BP values multiple times (typically 4 times/day) a day in different periods. Subjects were guided to rest for three minutes before taking the measurements in a sitting position. The study was approved by the Ethics Review Committee of the San-ei Medisys Company (#2019002SA). All methods were performed in accordance with the relevant guidelines and regulations. Informed consent was obtained from all the subjects.

Table 1. Individual fitting results based on GPR_set1 model coupled with intermittent calibration.

Sub.	Ag	Gender	Ref. BP Range		MAE		RMSE	
No.	Yrs.	F/M	SBP	DBP	SBP	DBP	SBP	DBP
1	45	F	119–90	87–63	5.01	6.13	6.29	7.23
2	43	M	131–100	98–66	4.60	4.60	5.70	5.91
3	27	M	141–101	88–65	6.23	4.60	7.62	5.57
4	50	M	173–129	125–97	6.20	4.31	7.84	5.34
5	46	F	110–92	79–61	3.36	4.01	4.28	4.13
6	50	M	127–94	89–69	5.63	6.83	7.30	8.95
7	44	M	138–108	110–78	5.23	4.64	6.43	5.49
8	25	F	119–93	81–56	4.49	3.92	5.53	4.98
9	46	M	136–91	99–72	5.70	5.37	7.77	6.84
10	47	F	141–111	97–78	4.30	3.35	5.14	4.16
11	29	F	144–88	93–56	6.22	4.06	8.24	4.78

2.1. Preprocessing

This research differentiates itself from the previous studies that tried to extract the morphological features of PPG waveforms as many as possible and then used the features selection methods to pick up the important ones to input into the regression models. We argue that the whole pipeline should take serious account of the problem that the deterioration of the dicrotic notch and the diastolic peak with aging and hypertension will bring uncertainty to the relevant features. This is also the reason why we highlight the *A2* aspect in our study.

On this account, the explanatory features set X consists of (1) the heart rate reading (*HR*) during the experiment; and (2) the general PPG waveform features, which are described as follows. Firstly, 6th order Butterworth IIR low pass zero-phase filter (f_c = 10 Hz) and a linear detrend process were applied to the PPG signals before the following procedure:

Firstly, the skewness values, which is a well-used index for the signal quality index (SQI) of PPG signal [20], of beat-by-beat PPG waveforms in each sample was calculated and the waveform with the maximum skewness values was picked, thereafter the sample with the maximum SQI less than 0.1 will be removed.

Secondly, the following features (Figure 1) of the waveform were calculated:

- *PPG* intensive ratio (*PIR*): $PIR = PPG_{peak}/PPG_{root}$, which is an index to reflect changes in the arterial diameter [21];

- Diastole time ratio (*DTR*): $DTR = t_{sys}/T$, where the t_{sys} is the time from the systole peak to the end of a PPG waveform, and T is the time length of the corresponding PPG waveform;
- *ri*: $ri = PPG_{peak}/PPG_{inf}$, where the PPG_{inf} is the inflection point of the PPG waveform between the dicrotic notch and the diastolic peak. This parameter is introduced given the disappearance of the diastolic peak, while the inflection point can be found. Given the fact that the PPG amplitudes of inflection point and the diastolic peak are similar, the *ri* is used as the alternative of the Reflection index [22];
- *A02*: Area of the 0–2 Hz band of the PPG waveform [6];
- *A25*: Area of the 2–5 Hz band of the PPG waveform [6].

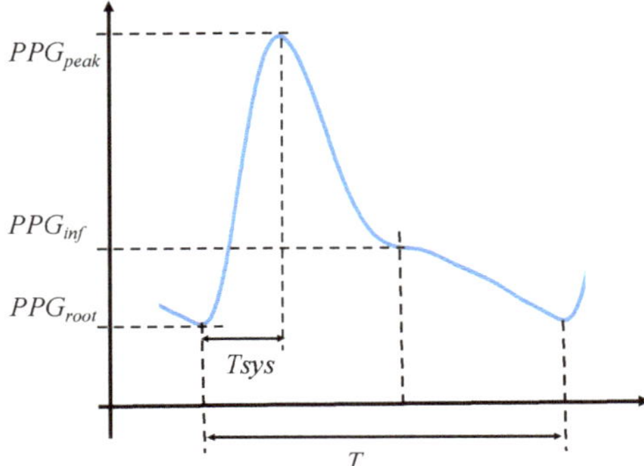

Figure 1. Illustration of PPG features. Inflection point of the PPG waveform between the dicrotic notch and the diastolic peak is used given that the disappearance of the diastolic peak may appear.

The five features above, which have been validated in the previous studies [4,6,2] were selected based on the underlying consideration that potential PPG features should avoid the identification of the dicrotic north and the diastole peak. Meanwhile, the H which has been proved to be informative in BP estimation [16,23], is used as another explanatory variable.

2.2. Regression Models

To respond to the *A1*, three regression models were used in this research. While the PLS corresponds to the linear assumption of the features (*X*) and the BP (*y*), the LWP corresponds to the assumption of local linearity defined by the close samples in the feature space. The Gaussian Process Regression can be viewed as a nonlinear regression by using Radial Basis kernel.

2.2.1. Partial Least Squares (PLS)

By projecting the input into new spaces, PLS can handle the problem with visible collinearity in the explanatory variables. In a problem with $X \in \Re^{N \times P}$ as the explanatory variables (*N*: the number of samples; *P* the number of the explanatory variables) and $y \in \Re^N$ as the dependent variable, the normalized *X* and *y* are expressed alternatively as follows:

$$X = TP^T + E,$$

$$y = Tq + f,$$

where the $T \in \Re^{N \times D}$ is the matrix that contains latent components in each column and $P \in \Re^{P \times D}$ is the matrix that contains the loading of each variable for the component in each row. $q \in \Re^D$ is the regression coefficient vector from latent variables to the output. *E* and *f* are the residuals.

The projection of X and y is done iteratively by maximizing the covariance of the $y^T t_i$, where t_i is the *i*th column of the T, taking the first two latent components as an example.

$$t_1 = Xw_1, \tag{3}$$

where the w_1 is the weight vector for the first component with the regularization that $\|w_1\| = 1$. By using the *Largrange* multiplier and least squares method, the w_1, the t_1, and the corresponding p_1 and q_1, which is the first row of P and the first element of q respectively, can be calculated. Similarly, the second component can be derived by subtracting the projection in the first component from the X and y,

$$X_2 = X - t_1 p_1^T, \tag{4}$$

$$y_2 = y - t_1 q_1, \tag{5}$$

and calculate the parameters by

$$w_2 = \frac{X_2^T y_2}{\|X_2^T y_2\|}, \tag{6}$$

$$P_2 = \frac{X_2^T t_2}{t_2^T t_2}, \tag{7}$$

$$q_2 = \frac{y_2^T t_2}{t_2^T t_2}. \tag{8}$$

The procedure is repeated when the i reaches the number D, whose value can be determined by using the cross-validated coefficient of determination (R_{cv}^2).

$$r_{cv}^2 = 1 - \frac{\sum_{i=1}^{N}(y^i - y_{cv}^i)^2}{\sum_{i=1}^{N}(y^i - y_A)^2}, \tag{9}$$

where the y^i and the y_{CV}^i are the true value and the estimate of the *i*th sample in the cross validation, respectively; the y_A is the mean of y. The i is set as 2 after the confirmation with r_{cv}^2.

2.2.2. LW-PLS

As a just-in-time method, the LW-PLS is conceptually different from the PLS. It separates the samples to a training dataset with $X_t \in \Re^{M \times P}$ and $y_t \in \Re^M$ and the new request $x_r \in \Re^P$, whose similarity will be compared with the training dataset and used to generate the diagonal similarity matrix.

$$U = \begin{bmatrix} u_r^1 & \cdots & 0 \\ \vdots & \ddots & \vdots \\ 0 & \cdots & u_r^M \end{bmatrix} \tag{10}$$

where the u_q^i is the similarity between the *i*th sample in X_t and the x_r. The X_t and y_t are adjusted based on the U.

$$X_0 = X_t - X_w, \tag{11}$$

$$y_0 = y_t - y_w. \tag{12}$$

The X_w and y_w are generated as follows.

$$X_w = \begin{bmatrix} 1 \\ 1 \\ \cdots \\ 1 \end{bmatrix} \begin{bmatrix} x_{w,1}, x_{w,2}, \cdots, x_{w,P} \end{bmatrix}, \tag{13}$$

and

$$y_w = \frac{\sum_{i=1}^{M} u_r^i y^i}{\sum_{i=1}^{N} u_r^i},\tag{14}$$

where,

$$x_{w,j} = \frac{\sum_{i=1}^{M} u_r^i x_j^i}{\sum_{i=1}^{N} u_r^i}.\tag{15}$$

With all this preparation the first component can be calculated.

$$w_1 = \frac{X_0^T U y_0}{\|X_0^T U y_0\|},\tag{16}$$

$$t_1 = X_0 w_1,\tag{17}$$

$$p_1 = \frac{X_0^T U t_1}{t_1^T U t_1},\tag{18}$$

$$q_1 = \frac{y_0^T U t_1}{t_1^T U t_1},\tag{19}$$

The p_1 and q_1 represent the loading vector and coefficient of the first component respectively. The parameters are determined iteratively when the predefined component number is met. The estimate for the request x_r is the summation of the $\hat{y}_a = \sum_{i=1}^{D} t_{a,i} q_{a,i} + y_w$

The distance index is crucial to the LW-PLS. Therefore, special care should be paid to its choice. In this research, it is considered that the *Mahalanobis* distance, which is the improved *Euclidian* distance to amend the problems of different scales and correlation between explanatory features with the definition below

$$d_m = \sqrt{(x_i - x_j)^T S^{-1}(x_i - x_j)}, \ (x_i \text{ and } x_j: \text{the samples}; S: \text{covariance matrix})\tag{20}$$

is suitable for this heterogeneous set of explanatory features consisting of continuous one of different scales.

2.2.3. Gaussian Process Regression

Gaussian Process Regression is fundamentally a kernel method, which defines the similarity of explanatory variables in terms of a kernel function. Instead of the explicit definition of the basic functions $\phi_i(x), (i = 1, 2, \ldots, l)$ and determination of the optimized weights vector w, where $y(x, w) = w^T \phi(x)$, the Gaussian Process defines a prior probability distribution over the basic functions. For a dataset with N samples, the equation becomes

$$y = \Phi w, \text{where } \Phi \text{ is the design matrix, } \Phi_{nk} = \phi_k(x_n).\tag{21}$$

By letting the prior distribution of the weights vector to be Gaussian, $p(w) \sim \mathcal{N}(w|0, \alpha^{-1})$ and then defining the kernel function $k(x, x')$, where

$$k(x, x') = \frac{1}{\alpha} \phi(x)^T \phi(x'),\tag{22}$$

The Gaussian Process can be defined by the kernel function, since $\mathbb{E}[y] = \Phi \mathbb{E}[w] =$

$$\mathbb{E}(y(x_n)y(x_m)) = k(x_n, x_m),\tag{23}$$

In this research, the kernel is the Radial Basis Function (scale = 1) contaminated with white noise (noise level = 1) for the standardized explanatory variables.

2.3. Calibration Schemes

The personality is reflected by incorporating the personal samples into the training dataset. Two kinds of calibration schemes were used. The first scheme (scheme_1) used the first 20% (typically 23 samples) of the samples in the training process; whereas the second scheme (scheme_2) used the first 4% and the 15th–18th, 35th–38th, 55th–58th, 75th–78th, and 95th–98th in the training process, which were roughly 20% of the individual samples as well. Of note, to prevent the leaking of future information, the BP values are predicted by using the previous calibration readings. That is, for example, the 19th-34th BP values are predicted by using the initial and the 15–18th calibration readings. These two schemes were used to examine the necessity of re-calibration in a relatively long-term scenario (*Q2*). Specifically, if the best results come from the the scheme_2, it is reasonable to conclude that the model should take the biorhythm, such as seasonality, into account and the re-calibration should be considered.

2.4. Generalized and Individual Models

To answer the question *Q1*, two training strategies were used in model construction. A generalized model trains the model as a generalized one by using the individual calibration samples as well as the samples from the other subjects. The other strategy is to construct an individual model, which was trained by using individual calibration samples solely. The two calibration schemes above were used in both models. The mean absolute error (MAE), as well as the mean and standard deviation of the regression error, were used as the main metrics; particularly, the MAE was used to determine the best model.

3. Results

By summarizing the one-month experiment, all of the 1287 samples collected from the 11 subjects, whose information is shown in Table 1, were used in the following modeling. The numbers of samples for the subjects are even, from 114 to 118.

3.1. Results of Generalized Models

In this study, the linear partial least square model (PLS), the local linear local-weighted partial least square model (LWPLS), and the Gaussian Process Regression model (GPR) were used to reflect different assumptions about this regression problem. The generalized model is trained in the manner of leave-one-out validation, by taking in all samples in the training dataset (10 subjects) merging the individual calibration samples (calibration scheme_1 and scheme_2, more detail in **Methods**). By summarizing the regression results of the testing samples, the mean absolute error (MAE) of systolic and diastolic BP is shown in Figure 2, from which it can be seen that none of the models can fit the personal well with MAE larger than 9 mmHg. Therefore, no further comparison of features was conducted for the generalized models.

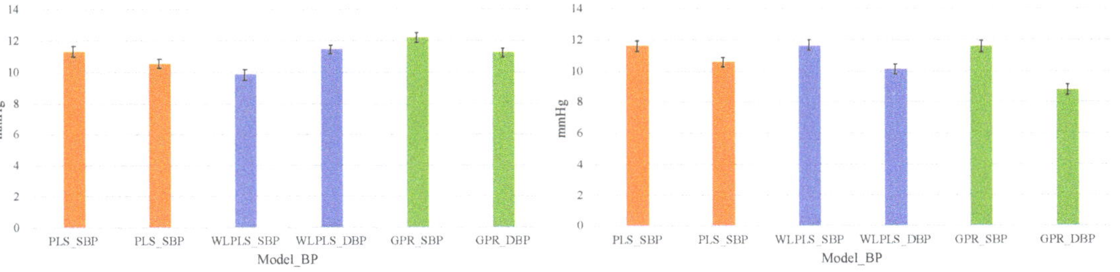

Figure 2. The mean absolute error (MAE) of systolic and diastolic BP of the three generalized models. The bar plot on the left corresponds to the results with scheme_1 and the bar plot on the right to the results with scheme_2. Sticks show the standard errors.

3.2. Comparison of Different Individual Models

Individual models were constructed by taking the calibration samples from the same subject for model construction and testing the samples remaining, whose results are summarized in Figure 3. As introduced in the **Methods** section, two calibration schemes were used. The left column corresponds to the calibration using the initial 20% samples (scheme_1), while the right column corresponds to the situation of re-calibrating every 20 samples (scheme_2).

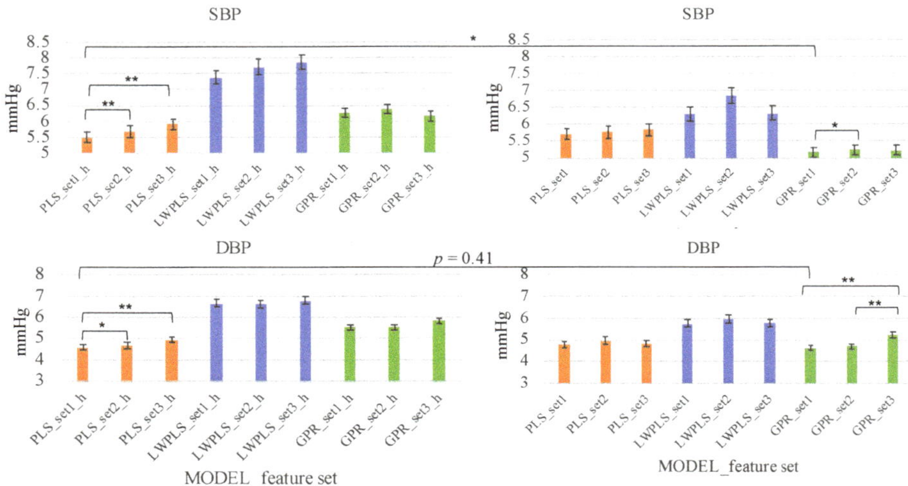

Figure 3. Comparison of three individual models using scheme_1 (left column) and scheme_2 (right column). Sticks show the standard errors. The setx means the combination of PPG features, where set1 is {*hr, a02, a25, ri*}, set2 is {*pir, dtr, hr, a02, a25, ri*}, set3 is {*hr, ri*}. * means the $p < 0.05$ and, ** means the $p < 0.001$. setx_h corresponds to the scheme_1 calibration.

By comparison with the generalized models, each model significantly outperforms its counterpart ($p < 0.001$). This suggests that the individual model is more appropriate building the BP regression model in the current stage.

The three models with two calibration schemes were examined by altering the combination of explanatory features. Therefore, there are three trials for each model, as shown in Figure 3. Inside each scheme, we only compare the model having the lowest MAE with others that have close MAEs by pair-wise student *t*-test and the models with lowest MAE from each scheme. For SBP, the best models are PLS_set1 for shceme_1 and GPR_set1 for scheme_2. Moreover, the GPR_set1 (scheme_2) is significantly smaller than PLS_set1 (shceme_1) and smaller than GPR_set2 (scheme_2); it is considered as the best model for SBP.

For DBP, being consistent with the situation of SBP, the best models are PLS_set1 (shceme_1) and GPR_set1 (scheme_2). However, no significant difference can be found from these two models ($p = 0.41$). Of note, regarding the individual model that uses scheme_2 calibration, linear interpolation that used two consecutive calibration windows has also been conducted to confirm the advantage of nonlinear regression using PPG features. For example, the averaged BP values of the two calibration windows of 15th–18th samples and the 35th–38th samples are used to interpolate the 19th–34th values. The MAEs of SBP and DBP are 5.96 and 4.72 mmHg, respectively. The accuracy of SBP is significantly lower than GRP_set1; whereas no significant difference can be seen for DBP. Combining these pieces of result, the feature combination set 1 is consistently the best. Moreover, the GPR coupled with the re-calibration scheme (scheme_2) has a significant low MAE for SBP (5.16 mmHg) and similar MAE for DBP (4.63 mmHg); it is considered as the best method for BP estimation.

3.3. Regression Model

Based on the results above, the GPR model with explanatory features of *hr*, *a02*, *a25*, and *ri* is used for both SBP and DBP regression. The results of the regression model are shown in Figure 4, from which it can be seen that the overall trend of the BP can be captured (Figure 4a,b). The decreasing trend of both the SBP and DBP may be attributed to the seasonal influence and can be observed from most of the subjects [24,25]. The mean and standard deviation of the fitting are

- SBP: -2.2 ± 6.2 mmHg; DBP: -2.0 ± 5.4 mmHg.

The individual fitting results are tabulated in Table 1, from which it can be confirmed that most of the subjects have SBP MAE values less than 6 mmHg, except for No. 3, 4, and 11, whose BP ranges were relatively wide.

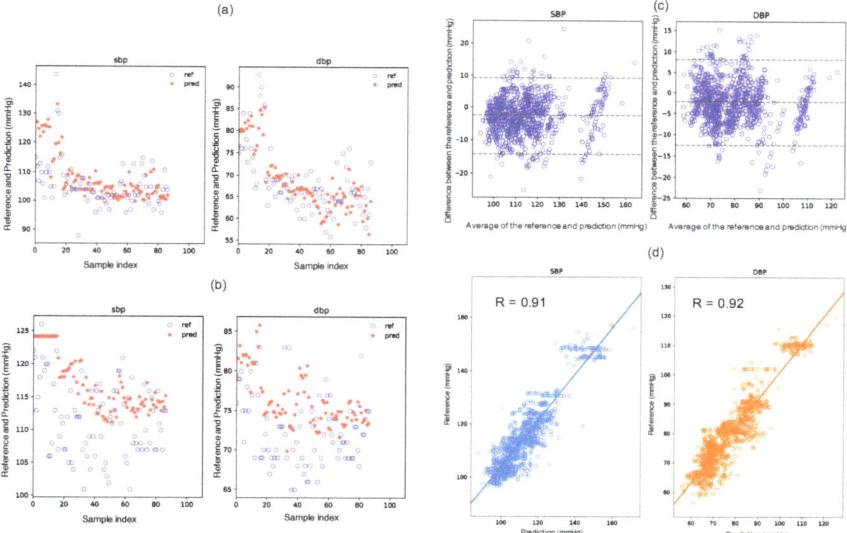

Figure 4. The results of the GPR_set1 (scheme_2) model. The fitting of the one-month samples of two randomly selected subjects can be seen in (**a**,**b**). The BA plot for and linear correlation can be found in (**c**,**d**).

4. Discussion

The observation that the generalized models are significantly inferior to individual models is consistent with the finding of Gašper et al. [26] based on the MIMIC data. This reconfirmation suggests that the samples (features and BP values) from different individuals follow different distributions. It also suggests that the generalizability of a BP model should be validated on a dataset within a certain period.

Although the ISO81060-2 uses the mean as a major criterion in model evaluation, MSE, which is more sensitive to the variance of the error, is often used in model selection [6,23]. In this study, the Mean errors of SBP and DBP with the PLS_set1 model are -1.46 and -1.53 mmHg, respectively. However, the correlation values are evidently lower as 0.82 and 0.89, respectively. Figure 5 shows the distributions of the predicting errors of PLS_set1 (left column) and GPR_set1 (right column). It can be confirmed in Figure 5 that the bias values of the PLS_set1 model are less; however, variances are higher than the GPR_set1 model.

The MAE also stands out in its insensitivity to the outliers compared with the mean square error, which is used pervasively in machine learning. Therefore, the MAE is used in model comparison in this study.

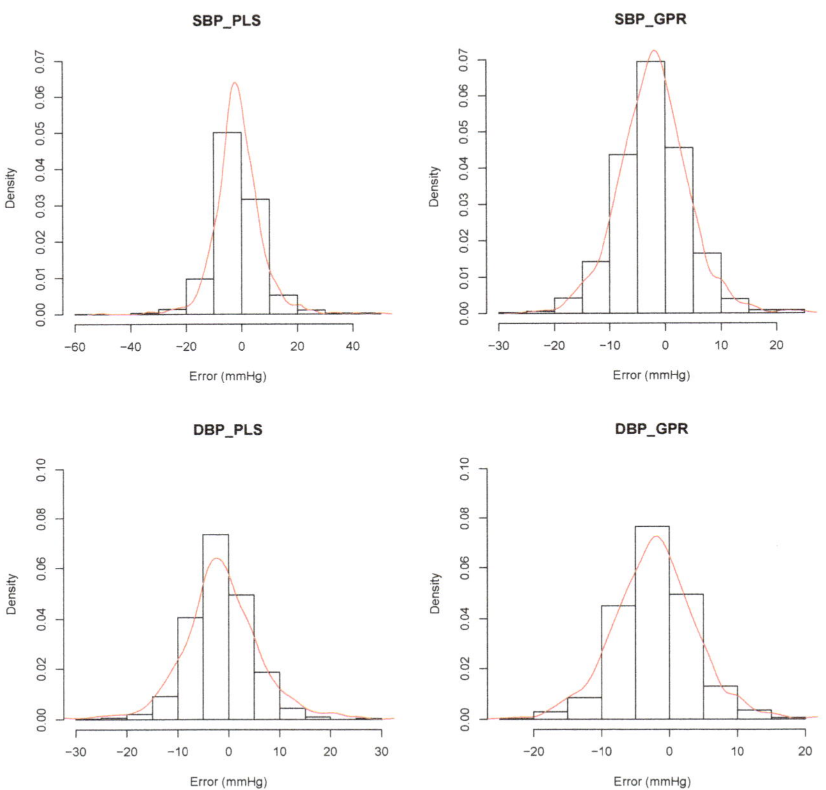

Figure 5. Distribution of predicting errors of the PLS_set1 (**left** column) adn GPR_set1 models (**right** column). Red curves show the approximation of probability density using Gaussian distribution.

The fact that the re-calibration scheme (scheme_2) surpasses the initial-calibration scheme (scheme_1) in systolic BP suggests that the model may only capture the fluctuation of the systolic BP within a couple of days and there is a component, periodically or not, with a longer period in the systolic BP change. On the other hand, the two schemes showed little difference in diastolic BP (lower row of Figure 4). This observation suggests that the diastolic BP is more stable than the systolic BP.

The necessary number of samples in a calibration window and the length of calibration-free interval between windows are important in planning a subsequent experiment with larger population and longer duration. By setting the target accuracy as 5 mmHg for both SBP and DBP, other settings of the scheme_2 have been tried, for example, two times of calibration at the 30th–40th, and the 70–80th samples result in significantly lower accuracy (MAE: 6.28 and 5.02 mmHg for SBP and DBP, respectively). Given that re-calibration too often may cause stress, an interval of 17–20 samples (4 to 5 days typically), and four samples of a calibration window (1 day typically), is suitable.

Another advantage of the re-calibration scheme is the swift initialization, which requires five measurements typically, whereas the initial-calibration scheme requires a lot period, 23 measurements typically, for initialization to have acceptable accuracy. Please note that the granularity of this conclusion is the number of days based on the setting of the experiments. Whether repeating the calibrating measurements in a shorter interval, e.g., several minutes, can get similar accuracy could be validated subsequently.

The necessity of an individual model could be explained partially by further plotting the distribution of similarity based on *Mahalanobis* distance (Figure 6). The intra-similarity of the explanatory variables from the same subject is higher than the inter-similarity but with partial overlapping. Since in kernel-based regression, the LWPLS and the GPR will take close samples in the explanatory variables space to predict the new sample, the overlapping samples may be detrimental for the prediction by deviating the predicting hyperplane.

Regarding the choice of the regression model, this insight is especially significant in the era of deep learning, which is fundamentally the data-driven approach. A shallow network model can give an accurate estimation for the dataset on hand given the extraordinary capability in twisting the hyperplane to fit the samples. However, given the strong personality underlying this problem, a few-shot learning scheme that relates the same PPG and relevant features from the same subject with the change of BP fluctuation may be a method worth trying at the present stage. Although the application of the fully data-driven approach, e.g., the end-to-end deep learning pipeline, onto this topic is still difficult for the lack of individual longitudinal data, the information hidden in the sequence of EEG waveforms could be used by recurrent neural network or attention structure in the near future with further accumulation of data. In this regard, the construction of long term individual datasets should be the next key work.

An interesting study that significantly increases the training samples by extending the measurement duration (30 min) and then uses the BiLSTM model to predict the BP values hints that the deep learning structures should be carefully used for this topic. The main conclusion of the leave-one-subject-out validation that the model pretrained with population samples and fine-tuned with individual data attains the best performance may be partially attributed to a more informative feature space constructed by ECG, PPG, and BCG signal than using the PPG sensor solely. That is, the general relation between the features extracted from multiple sensors and the BP values could be captured and used in transfer learning. Although the physiological measurement is complicated, given that the simultaneous measurement of ECG, PPG, and BCG is feasible nowadays, researchers and developers are therefore recommended to decide on the sources of signal based on the target application. However, if the validation duration extended to multiple days, the significant decrease in predicting error suggests the influence of biorhythm [27].

The results of this paper are consistent with the results of Chowdhury et al. [6] by showing that Gaussian Process Regression can attain accurate BP estimation. Additionally, this research distinguishes itself by using the features that do not require the identification of the dicrotic notch and the diastolic peak and validating using a long-term dataset to highlight the necessity of the individual model.

This research emphasizes the importance of feature selection. There is no reason to replace the features which have been validated in previous studies with the automatically generated features using deep learning models in such a data-insufficient problem. At least, at the current stage, the domain knowledge is as important as the improvement of machine learning models.

This research shows the possibility of building an accurate BP estimation model by using the general features of PPG signal and HR. Moreover, given that the disappearance of diastolic peak is inevitable in aged people, who are the targeted population of the cuffless method, the general features could be a good alternative in modeling.

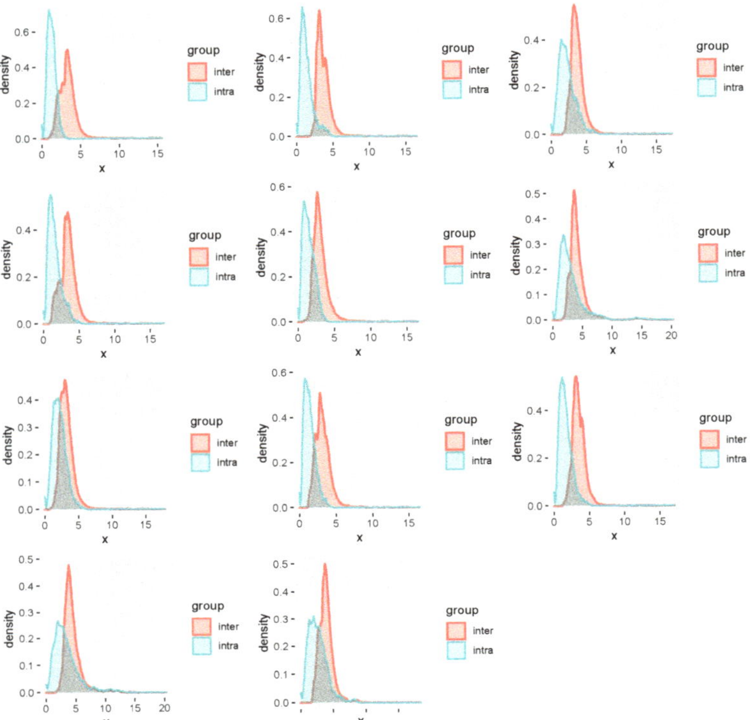

Figure 6. Distribution of similarity. Each subfigure corresponds to one subject. In each subfigure, t[...] histogram in red shows the distribution of similarity of samples from the same subject; whereas t[...] one in blue shows the distribution of similarity with other subjects.

5. Conclusions

This paper, from the perspectives of data science and biomedical engineering, provid[...] insightful results that are important for successive cuffless blood pressure studies. The ma[...] results of this research are (1) in the current stage of data insufficiency, an individu[...] nonlinear regression model with intermittent calibration scheme is an ideal alternati[...] for long-term use; (2) the general waveform features are used given the disappearance [...] morphological features caused by aging. Given the good fitting results of the nonline[...] regression model, the general waveform features are informative in this regression probl[...] and therefore can be used in the future studies.

Author Contributions: T.T., M.T. and M.H. conceived the experiments; T.T., M.T. and M.H. conduc[...] the experiments; M.H., N.O., S.K. and Z.C. designed the model; M.H. and Z.C. wrote the paper; M[...] Z.C. and K.K. revised the model and code; M.H., T.T., N.O., W.C., M.A.-U.-A. and S.K. revised [...] paper and helped with interpretation and discussion; T.T., M.T. and M.H. supervised model des[...] and experiments. All authors have read and agreed to the published version of the manuscript.

Funding: This research and development work was supported by the KYOTO Industrial Sup[...] Organization 21, the Kigyonomori Grant, the Grant-in-Aid for Early-Career Scientists (Kake[...] (#20K19923), the Grant-in-Aid Scientific Research (C) (Kakenhi) (#17K01440, #21K12760), and [...] Japan Agency for Medical Research and Development (grant no. s20dk0310111).

Institutional Review Board Statement: The study was approved by the Ethics Review Commi[...] of the San-ei Medisys Company (#2019002SA).

Informed Consent Statement: Informed consent was obtained from all subjects involved in the study.

Data Availability Statement: All data analyzed during the current study are available from the corresponding author on reasonable request.

Acknowledgments: The authors would like to thank Nobuhiro Nichimura from San-Eki Medisys Co. Ltd. for the data collection.

Conflicts of Interest: The authors declare no conflict of interest.

References

1. Chen, W.; Kobayashi, T.; Ichikawa, S.; Takeuchi, Y.; Togawa, T. Continuous Estimation of Systolic Blood Pressure Using the Pulse Arrival Time and Intermittent Calibration. *Med. Biol. Eng. Comput.* **2000**, *38*, 569–574. [CrossRef] [PubMed]
2. Tang, Z.; Tamura, T.; Sekine, M.; Huang, M.; Chen, W.; Yoshida, M.; Sakatani, K.; Kobayashi, H.; Kanaya, S. A Chair–Based Unobtrusive Cuffless Blood Pressure Monitoring System Based on Pulse Arrival Time. *IEEE J. Biomed. Health Inform.* **2017**, *21*, 1194–1205. [CrossRef] [PubMed]
3. Xing, X.; Ma, Z.; Zhang, M.; Zhou, Y.; Dong, W.; Song, M. An Unobtrusive and Calibration-Free Blood Pressure Estimation Method Using Photoplethysmography and Biometrics. *Sci. Rep.* **2019**, *9*, 1–8. [CrossRef] [PubMed]
4. Shin, H.; Min, S.D. Feasibility study for the non-invasive blood pressure estimation based on PPG morphology: Normotensive subject study. *Biomed. Eng. Online* **2017**, *161*, 1–14. [CrossRef] [PubMed]
5. Chandrasekhar, A.; Kim, C.S.; Naji, M.; Natarajan, K.; Hahn, J.O.; Mukkamala, R. Smartphone-Based Blood Pressure Monitoring via the Oscillometric Finger-Pressing Method. *Sci. Transl. Med.* **2018**, *10*, 1–12. [CrossRef] [PubMed]
6. Chowdhury, M.H.; Shuzan, M.N.I.; Chowdhury, M.E.; Mahbub, Z.B.; Uddin, M.M.; Khandakar, A.; Reaz, M.B.I. Estimating Blood Pressure from the Photoplethysmogram Signal and Demographic Features Using Machine Learning Techniques. *Sensors* **2020**, *20*, 3127. [CrossRef] [PubMed]
7. Luo, H.; Yang, D.; Barszczyk, A.; Vempala, N.; Wei, J.; Wu, S.J.; Zheng, P.P.; Fu, G.; Lee, K.; Feng, Z.P. Smartphone-based blood pressure measurement using transdermal optical imaging technology. *Circ. Cardiovasc. Imaging* **2019**, *128*, 1–10. [CrossRef] [PubMed]
8. Foo, J.Y.A.; Lim, C.S.; Wang, P. Evaluation of Blood Pressure Changes Using Vascular Transit Time. *Physiol. Meas.* **2006**, *27*, 685–694. [CrossRef]
9. Elgendi, M.; Fletcher, R.; Liang, Y.; Howard, N.; Lovell, N.H.; Abbott, D.; Lim, K.; Ward, R. The Use of Photoplethysmography for Assessing Hypertension. *Npj Digit. Med.* **2019**, *2*, 1–11. [CrossRef]
10. Maeda, Y.; Sekine, M.; Tamura, T. The advantages of wearable green reflected photoplethysmography. *J. Med. Syst.* **2011**, *355*, 829–834. [CrossRef] [PubMed]
11. Allen, J. Photoplethysmography and its application in clinical physiological measurement. *Physiol. Meas.* **2007**, *283*, 1–39. [CrossRef] [PubMed]
12. Wisely, N.A.; Cook, L.B. Arterial flow waveforms from pulse oximetry compared with measured Doppler flow waveforms. *Anaesthesia* **2001**, *566*, 556–561. [CrossRef] [PubMed]
13. O'Rourke, M.F. Vascular impedance in studies of arterial and cardiac function. *Physiol. Rev.* **1982**, *622*, 570–623. [CrossRef] [PubMed]
14. Allen, J.; Murray, A. Age-related changes in the characteristics of the photoplethysmographic pulse shape at various body sites. *Physiol. Meas.* **2003**, *242*, 297–307. [CrossRef] [PubMed]
15. Makino, M.; Hayashi, H.; Takezawa, H.; Hirai, M.; Saito, H.; Ebihara, S. Circadian Rhythms of Cardiovascular Functions Are Modulated by the Baroreflex and the Autonomic Nervous System in the Rat. *Circulation* **1997**, *96*, 1667–1674. [CrossRef]
16. Chen, Y.; Shi, S.; Liu, Y.K.; Huang, S.L.; Ma, T. Cuffless Blood-Pressure Estimation Method Using a Heart-Rate Variability-Derived Parameter. *Physiol. Meas.* **2018**, *39*, 95002. [CrossRef]
17. Esmaili, A.; Kachuee, M.; Shabany, M. Nonlinear Cuffless Blood Pressure Estimation of Healthy Subjects Using Pulse Transit Time and Arrival Time. *IEEE Trans. Instrum. Meas.* **2017**, *66*, 3299–3308. [CrossRef]
18. Miki, K.; Yoshimoto, M. Exercise-Induced Modulation of Baroreflex Control of Sympathetic Nerve Activity. *Front. Neurosci.* **2018**, *12*, 1–6. [CrossRef] [PubMed]
19. Kario, K.; Saito, K.; Sato, K.; Hamasaki, H.; Suwa, H.; Okura, A.; Hoshide, S. Validation of the A&D BP UA-1200BLE device for home blood pressure measurement according to the ISO 81060–2: 2013 standard. *Blood Press. Monit.* **2018**, *23*, 312–314.
20. Liang, Y.; Elgendi, M.; Chen, Z.; Ward, R. Analysis: An Optimal Filter for Short Photoplethysmogram Signals. *Sci. Data* **2018**, *5*, 1–12. [CrossRef] [PubMed]
21. Ding, X.; Zhang, Y.; Liu, J.; Dai, W.; Tsang, H.K. Continuous Cuffless Blood Pressure Estimation Using Pulse Transit Time and Photoplethysmogram Intensity Ratio. *IEEE Trans. Biomed. Eng.* **2016**, *635*, 964–972. [CrossRef] [PubMed]
22. Yousef, Q.; Reaz, M.B.I.; Ali, M.A.M. The Analysis of PPG Morphology: Investigating the Effects of Aging on Arterial Compliance. *Meas. Sci. Rev.* **2012**, *12*, 266–271. [CrossRef]
23. Hayase, T. Blood Pressure Estimation Based on Pulse Rate Variation in a Certain Period. *Sci. Rep.* **2020**, *10*, 1–14. [CrossRef]

24. Kollias, A.; Kyriakoulis, K.G.; Stambolliu, E.; Ntineri, A.; Anagnostopoulos, I.; Stergiou, G.S. Seasonal blood pressure variation assessed by different measurement methods: Systematic review and meta-analysis. *J. Hypertens.* **2020**, *28*, 791–798. [CrossRef] [PubMed]
25. Modesti, P.A.; Morabito, M.; Massetti, L.; Rapi, S.; Orlandini, S.; Mancia, G.; Gensini, G.F.; Parati, G. Seasonal blood pressure changes: An independent relationship with temperature and daylight hours. *Hypertension* **2013**, *61*, 908–914. [CrossRef]
26. Slapničar, G.; Mlakar, N.; Luštrek, M. Blood Pressure Estimation from Photoplethysmogram Using a Spectro-Temporal Deep Neural Network. *Sensors* **2019**, *19*, 3420. [CrossRef] [PubMed]
27. Lee, D.; Kwon, H.; Son, D.; Eom, H.; Park, C.; Lim, Y.; Seo, C.; Park, K. Beat-to-beat continuous blood pressure estimation using bidirectional long short-term memory network. *Sensors* **2021**, *21*, 96. [CrossRef]

Article

SMCKAT, a Sequential Multi-Dimensional CNV Kernel-Based Association Test

Nastaran Maus Esfahani [1,*,†] , Daniel Catchpoole [1,2,†] and Paul J. Kennedy [1,†]

1   Australian Artificial Intelligence Institute, University of Technology Sydney, Sydney 2007, Australia; Daniel.Catchpoole@uts.edu.au (D.C.); Paul.Kennedy@uts.edu.au (P.J.K.)
2   The Tumour Bank, The Children's Hospital at Westmead, Sydney 2145, Australia
*   Correspondence: nastaran.mausesfahani@student.uts.edu.com
†   These authors contributed equally to this work.

Abstract: Copy number variants (CNVs) are the most common form of structural genetic variation, reflecting the gain or loss of DNA segments compared with a reference genome. Studies have identified CNV association with different diseases. However, the association between the sequential order of CNVs and disease-related traits has not been studied, to our knowledge, and it is still unclear that CNVs function individually or whether they work in coordination with other CNVs to manifest a disease or trait. Consequently, we propose the first such method to test the association between the sequential order of CNVs and diseases. Our sequential multi-dimensional CNV kernel-based association test (SMCKAT) consists of three parts: (1) a single CNV group kernel measuring the similarity between two groups of CNVs; (2) a whole genome group kernel that aggregates several single group kernels to summarize the similarity between CNV groups in a single chromosome or the whole genome; and (3) an association test between the CNV sequential order and disease-related traits using a random effect model. We evaluate SMCKAT on CNV data sets exhibiting rare or common CNVs, demonstrating that it can detect specific biologically relevant chromosomal regions supported by the biomedical literature. We compare the performance of SMCKAT with MCKAT, a multi-dimensional kernel association test. Based on the results, SMCKAT can detect more specific chromosomal regions compared with MCKAT that not only have CNV characteristics, but the CNV order on them are significantly associated with the disease-related trait.

Keywords: genetic variation; copy number variants; disease-related traits; sequential order; association test

1. Introduction

Genetically speaking, all humans are 99.9 percent the same and the 0.1 percent that makes us all unique is called genetic variation [1]. Genetic variation has two main forms: structural alteration and sequence variation. Copy number variant (CNV) and DNA sequence variation are the most common form of structural alteration and sequence variation in the human genome, respectively [2].

A sequence variation or single nucleotide polymorphism (SNP) represents a difference in a single nucleotide. For example, a SNP may replace the nucleotide cytosine with the nucleotide thymine in a certain stretch of DNA. SNPs are classified into two major types based on the gene region they fall within: coding region and non-coding region. SNPs within a coding sequence do not necessarily change the amino acid sequence of the protein that is produced, due to the degeneracy of the genetic code. SNPs in the coding region are of two types: nonsynonymous and synonymous SNPs. Nonsynonymous SNPs change the amino acid sequence of the protein, while synonymous SNPs do not affect the amino acid sequence of the protein. SNPs do not usually function individually, rather, they work in coordination with other SNPs to manifest a disease or trait [3]. Therefore, many sequence studies have been done to test the association between SNPs and disease or traits.

Citation: Maus Esfahani, N.; Catchpoole, D.; Kennedy, P.J. SMCKAT, a Sequential Multi-Dimensional Kernel-Based CNV Association Test. *Life* **2021**, *11*, 1302. https://doi.org/10.3390/life11121302

Academic Editors: Md. Altaf-Ul-Amin, Tao Huang, Shigehiko Kanaya, Naoaki Ono, Ming Huang

Received: 15 September 2021
Accepted: 23 November 2021
Published: 26 November 2021

Publisher's Note: MDPI stays neutral with regard to jurisdictional claims in published maps and institutional affiliations.

A copy number variant is the gain or loss of DNA segments in the genome ranging in size from one kilobase to several megabases. CNVs are described by three characteristics type, chromosomal position, and dosage [4]. The type of CNV is either amplification or deletion. The chromosomal position of the CNV is described by the start and end position of the CNV in the chromosome. The dosage represents the total number of copies of the CNV with a value less than two relating to deletion and greater than two indicating amplification. Besides, CNVs have phenotypic heterogeneity effects. This means that different CNV types and dosages at the same position in the chromosome can have a different impact. It is reported in biological studies that CNVs are distributed non-randomly in the genomes, in particular they tend to be located close to telomeres and centromeres [5]. However, it is still unclear if there is any specific pattern in the sequential order of CNVs that may lead to a disease or trait.

Association studies have determined that genetic variations, both CNVs and SNPs are associated with diseases or traits. So, understanding the relationship between genetic variation and disease may provide important insights into genetic causes, leading to effective means in preventing and treating the diseases. While there are lots of computational association studies that have investigated the association between SNPs and diseases or traits, methods for studying CNVs are underdeveloped due to the multi-dimensional characteristics of the CNVs.

The CNV kernel association test (CKAT) [6], copy number profile curve-based association test (CONCUR) [7] and multi-dimensional copy number variant kernel association test (MCKAT) [8] are a few existing computational kernel based methods that have studied the association between CNVs and diseases. In these studies, different kernels are proposed to measure the similarity between CNV profiles with respect to CNV characteristics. Then the similarity between CNV profiles is compared with those in disease-related trait states to identify any potential association between CNVs and the disease. Among them, our previous method, the MCKAT, is the only method that has incorporated all multidimensional characteristics of CNVs in testing the association between CNVs and disease or trait. The MCKAT calculates the p-value of the association test analytically, which is computationally efficient and flexible for CNV association analysis for both rare and common CNV types, as demonstrated in numerical studies. However, neither MCKAT nor other methods consider the CNV sequential order in testing the association between CNVs and disease-related traits.

Starting from MCKAT, we propose a sequential multi-dimensional CNV kernel-based association test (SMCKAT) for investigating the association between CNVs and disease or traits. SMCKAT is not only utilizing all multi-dimensional characteristics of CNVs but also the sequential order of CNVs in testing the association between CNVs and disease traits. Based on the results, SMCKAT is applicable on both rare and common datasets and capable of identifying hot-spots on the genome where both CNV characteristics and the CNV sequential order are significantly associated with disease or traits.

The rest of this paper is as follows. Section 2 presents the method and material. Section 3 contains simulation studies. Section 4 shows the results of the real data application. Section 5 presents the discussion, and finally Section 6 concludes the work.

2. Method and Materials

We design a sequential multi-dimensional kernel framework capable of measuring similarity between CNV profiles utilizing all CNV characteristics and the CNV sequential order. It contains two kernels. The first kernel, the pair group kernel, measures similarity between two groups of CNVs at the same ordinal position of two CNV profiles. It contains three sub-kernels. Each sub-kernel is responsible for measuring the similarity between two CNVs with respect to one of the three CNV characteristics. The second kernel, the whole genome group kernel, aggregates the similarity between every possible CNV pair group to measure the total similarity between the CNV profiles of the subjects. Finally, the association between CNV sequential order across a chromosome and disease-related traits is tested by comparing the similarity in CNV profiles to that in the trait using an association test.

2.1. Pair CNV Group Kernel

Let X denote a single CNV which is defined by four characteristics as $X = (X^{(1)}, X^{(2)}, X^{(3)}, X^{(4)})$ where $X^{(1)}$ and $X^{(2)}$ are the CNV starting and ending position on the chromosome, $X^{(3)}$ is the CNV type, and $X^{(4)}$ is the CNV dosage. First, we generate the CNV profile R for subject i with l CNVs as $R_i = (X_1^i, X_2^i, \ldots, X_{l_i}^i)$ where CNVs are sorted based on their chromosomal position. Secondly, we extract a CNV group of size n out of the CNV profile as $G_i = (X_m^i, X_{m+1}^i, \ldots, X_{m+n}^i)$ where n is the group size that can take any value between 1 and l, the number of existing CNVs in a CNV profile as is shown in Figure 1.

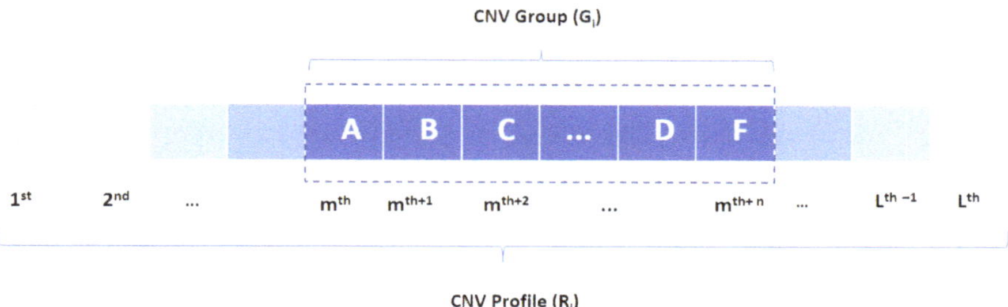

Figure 1. Generating CNV profile R_i where CNVs are sorted with respect to their chromosomal position. A, B,..., and F are arbitrary CNVs at $m^{th}, m^{th+1}, \ldots,$ and m^{th+n} positions and G_i is a group of CNVs of size n.

We propose a pair CNV group kernel, K_{PG}, to measure the similarity between two CNV groups of size n, G_i and G_j, in two CNV profiles. First, K_{PG} aligns each CNV in the G_i with its relevant CNV in the G_j with respect to their position to generate n CNV pairs as is shown in Figure 2.

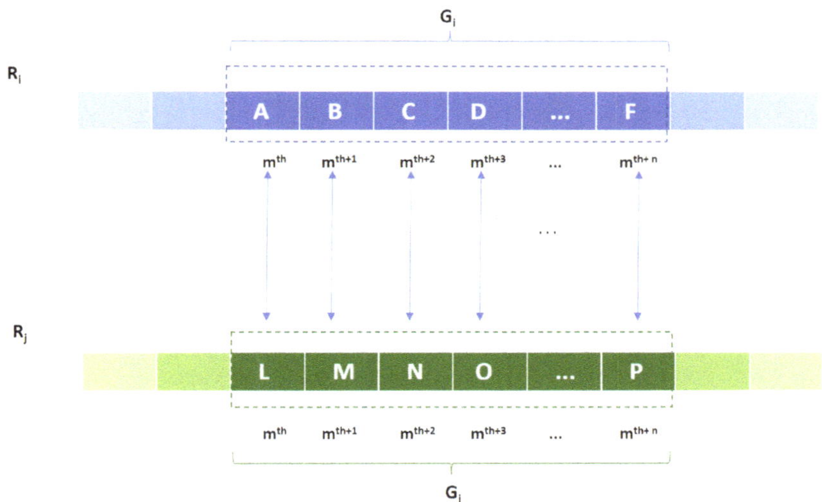

Figure 2. Aligning CNVs within two CNV groups of size n, G_i and G_j, to generate n CNV pairs.

Then, K_{PG} measures the similarity between each CNV pair using the single pair CNV kernel, K_S, we proposed in [8]. K_S measures the similarity between a CNV pair by three sub-kernels considering all CNV features including chromosomal position, type and dosage. Finally, K_{PG} averages the similarities calculated by K_S between all generated CNV pairs to measure the similarity between two CNV groups, G_i and G_j, as

$$K_{PG}(G_i, G_j) = \sum_{m=1}^{n} \frac{K_s(X_m^i, X_m^j)}{n} \tag{1}$$

where K_s is defined as

$$K_s\left(X_m^i, X_m^j\right) = \left[\frac{Intersection\left(\left(X_m^{i(1)}, X_m^{i(2)}\right), \left(X_m^{j(1)}, X_m^{j(2)}\right)\right)}{Union\left(\left(X_m^{i(1)}, X_m^{i(2)}\right), \left(X_m^{j(1)}, X_m^{j(2)}\right)\right)}\right] \times$$
$$\left[\frac{\left(X_m^{i(3)} == X_m^{j(3)}\right) + 1}{2}\right] \times \left[\frac{1}{2^{\left|DR\left(X_m^{i(4)}\right) - DR\left(X_m^{j(4)}\right)\right|}}\right] \tag{2}$$

and the first term measures the mutual presence of a CNV with a specific start and end position by dividing the size of the intersection of two CNVs to their union size. The intersection function calculates the length of the chromosomal region that belongs to both CNVs. Similarly, the union function calculates the length of the chromosomal region that consists of both regions that belong to the first CNV and to the second CNV. The second term compares the CNV type of two CNVs to calculate the similarity between them. The third term measures the similarity between two CNVs with respect to their dosage. The DR is the difference from the reference function we proposed in [8] a $DR(dosage) = |dosage - 2|$. DR measures the difference between a CNV dosage and the reference dosage value 2.

2.2. Whole Genome CNV Group Kernel

First, we create a window of size n. We slide this window across the CNV profile R_i as is shown in Figure 3 to extract all possible CNV groups of size n as $P_i = (G_1^i, \ldots, G_{p_i}^i)$, where CNV groups are sorted based on their position and p_i is the number of extracted CNV groups for the CNV profile R_i. Similarly, we have another CNV group series $P_j = (G_1^j, \ldots, G_{q_j}^j)$ for CNV profile R_j.

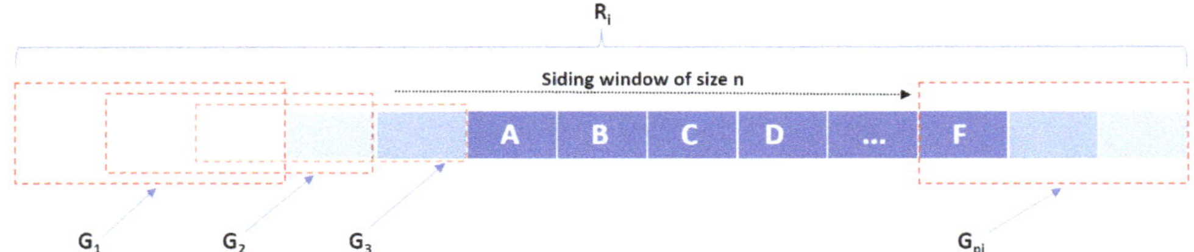

Figure 3. Sliding window of size n across CNV profile to extract CNV groups of size n.

Then, we propose the whole genome CNV group kernel, K_{WG}, to measure the similarity between two CNV group series P_i and P_j as

$$K_{WG}(P_i, P_j) = \begin{cases} 0 & \text{if } p_i \times q_i = 0 \\ \sum_{z=1}^{Max(p_i, q_i)} Max(K_{PG}(G_z^i, G_{z-1}^j), \\ K_{PG}(G_z^i, G_z^j), K_{PG}(G_z^i, G_{z+1}^j)) & \text{if } p_i \times q_i \neq 0 \end{cases}$$

where $K_{PG}(.,.)$ is the pair CNV group kernel from (1). K_{WG} measures the similarity between the pair CNV groups of the same position and aggregates these similarities to calculate the similarity in two CNV group series. The second maximum operation in the definition of K_{WG} searches for the best group-to-group correspondence of the highest similarity to align CNV groups in two CNV group series as is shown in Figure 4.

The kernel-based association test described in the following section, requires a kernel similarity matrix K. K is a $d \times d$ matrix, where $K_{ij} = K_{WG}(P_i, P_j)$ and d is the number of existing CNV profiles. K_{ij} expresses the similarity between CNV profile i and j measured by K_{WG}.

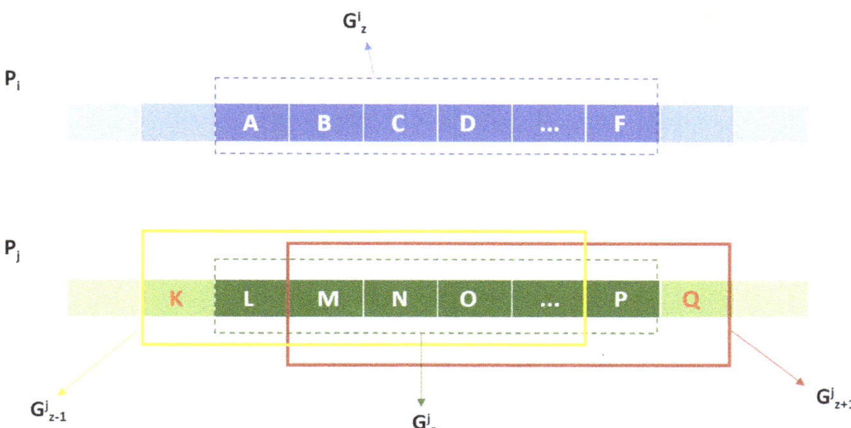

Figure 4. Aligning G^i_z to the best group-to-group correspondence of the highest similarity among G^j_{z-1}, G^j_z and G^j_{z+1}.

2.3. Kernel-Based Association Test

We use the following logistic regression model to test the association between CNV sequential order and a disease related trait

$$logit[Pr(y_i = 1)] = \beta_0 + Z\beta + f(P_i) \tag{4}$$

where y_i is the status of the disease related trait with $y_i = 1$ denoting the existence of the trait and $y_i = 0$ denoting otherwise, and $i = 1, 2, \ldots, d$ indexing the CNV profiles, and Z is the covariate matrix including information such as age and gender. P_i is the CNV group series of the profile R_i as explained previously. $f(\cdot)$ is a function spanned by the whole genome CNV group kernel $K_{WG}(\cdot, \cdot)$. According to equation (4), the hypothesis of no association between the CNV sequential order and the existence of a disease related trait can be tested as $H_0 : f(\cdot) = 0$. To test this, one way is to treat the $f(\cdot)$ as a random effect vector which is distributed as $N(0, \tau K)$, where $\tau \geq 0$ and K is the $d \times d$ similarity matrix, treated as covariance matrix of the random effect, generated by K_{WG} as defined in [6]. Liu et al. [9] has shown that testing $H_0 : f(\cdot)$ is equivalent to testing $H_0 : \tau = 0$ in the logistic mixed effect model. Moreover, τ is a variance component parameter in the logistic mixed effect model, which can be tested using a restricted maximum likelihood-based score test [9,10].

We use the following score test statistic where $\hat{y}$ is estimated under the null model $logit[Pr(y_i = 1)] = \beta_0 + Z\beta$ and K is the similarity matrix explained in the previous section.

$$Q = (y - \hat{y})' K (y - \hat{y}) \tag{5}$$

Then, we used the Davies method [11] as implemented in the CKAT R package [6] to calculate the p-value of the proposed kernel based association test. The SMCKAT workflow is summarized in Figure 5

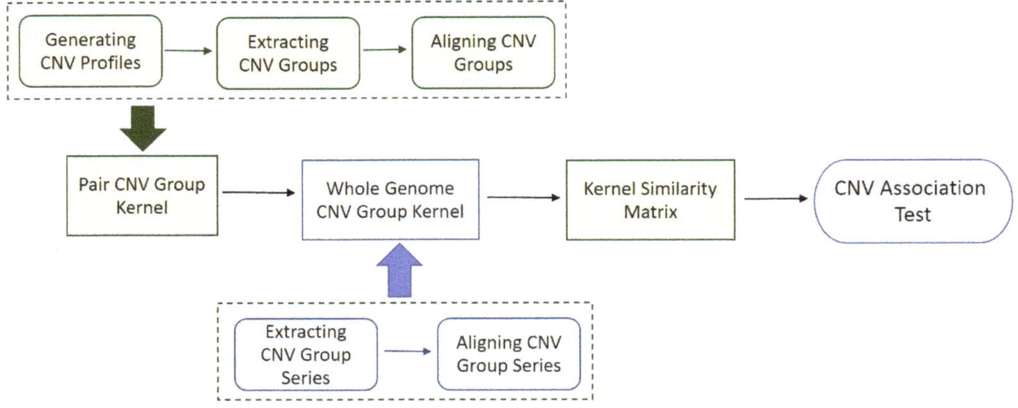

Figure 5. SMCKAT workflow diagram.

2.4. Common and Rare CNV Data

Biologists generally assign CNVs to one of two major types, depending on the length of the affected chromosomal region and occurrence frequency: copy number polymorphism (CNPs) and rare variants [4]. CNPs are widespread in the general population, with an average occurrence frequency greater than one percent while rare variants are much longer than CNPs, ranging from hundreds of thousands of base pairs to over one million base pairs.

We apply SMCKAT on both rare and common CNV public domain genome sequencing data sets to evaluate the performance on both CNV types. The two CNV data sets used in this study are from individuals with rhabdomyosarcoma (RMS) cancer and autism spectrum disorder (ASD). The RMS data set [12] contains the common CNVs for 44 subjects while the ASD data set [13] has the rare CNVs of 588 subjects. In both data sets, each CNV is presented by chromosomal position, type, and dosage.

3. Simulation Studies

We conducted simulations to evaluate the performance of SMCKAT and ensure that it can properly handle type I and II errors, as well as having relatively high power detecting existing associations. Besides SMCKAT, the MCKAT and CKAT are also studied. We conduct our simulation studies under two main scenarios. In the first scenario, we evaluate the performance of the SMCAKT on the rare CNV data. In the second scenario, we evaluate the performance of the SMCKAT on the common CNV data.

We use the ASD dataset and the RMS dataset in the first and second simulation scenarios, respectively. These datasets are studied in the real data analysis and further details regarding them are shared in the Section 4. We simulated 10^5 datasets for each simulation scenario.

The ASD dataset has the same dosage value for all deletions and similarly the same dosage value for all amplifications. Therefore, we randomly generate other values for the CNV dosage to conduct our simulation studies and investigate the dosage effect in identifying existing associations. The simulated dosage value can take 0 or 1 for deletion types and $3, 4, \ldots, 7$ for amplification types. We use equal probabilities when generating random dosage values for deletion and amplification, 0.5 and 0.2, respectively.

A case-control phenotype is generated for both SMCKAT and MCKAT from following logistic model that we proposed in [8],

$$logit(Pr(Y_i = 1)) = \beta_0 + \sum_{j=1}^{m_i} \beta_j^{Len}(X_{ij}^{(2)} - X_{ij}^{(1)}) + \sum_{j=1}^{m_i}(\beta_j^{Del}I[X_{ij}^{(3)} = 1] + \beta_j^{Amp}I[X_{ij}^{(3)} = 3])$$

$$+ \sum_{j=1}^{m_i} \beta_j^{Dsg}|X_{ij}^{(4)} - 2| + \sum_{j=1}^{m_i} \beta_j^{Len*Del*Dsg}(X_{ij}^{(2)} - X_{ij}^{(1)}) \times I[X_{ij}^{(3)} = 1] \times X_{ij}^{(4)} \tag{6}$$

$$+ \sum_{j=1}^{m_i} \beta_j^{Len*Amp*Dsg}(X_{ij}^{(2)} - X_{ij}^{(1)}) \times I[X_{ij}^{(3)} = 3] \times X_{ij}^{(4)}$$

where $X_{ij} = (X_{ij}^{(1)}, X_{ij}^{(2)}, X_{ij}^{(3)}, X_{ij}^{(4)})$ is the jth CNV of the ith individual as defined previously. β_0 corresponds to a baseline disease rate. β_j^{Len} controls the effect of chromosomal position, and β_j^{Del} and β_j^{Dup} are the log ratio of a CNV j for being deletion versus amplification and vice versa. β_j^{Del} and β_j^{Dup} share the same values but different signs. $\beta_j^{Len*Amp*Dsg}$ and $\beta_j^{Len*Del*Dsg}$ allow the effect of the chromosomal position and CNV type to differ by dosage in CNV j.

After generating phenotypes for SMCKAT and MCKAT, we use following logistic model that is proposed in [6] to generate the phenotypes under CKAT method:

$$logit(\pi_i) = \beta_0 + \sum_{j=1}^{m_i}(\beta_j^{Del}I[X_{ij}^{(2)} = 1] + \beta_j^{Dup}I[X_{ij}^{(2)} = 3])X_{ij}^{(1)} \tag{7}$$

where $X_{ij} = (X_{ij}^{(1)}, X_{ij}^{(2)})$ is the jth CNV of ith subject, $\pi_i = Pr(Y_i = 1)$, β_0 is the prevalence rate of the disease, and β_j^{Dup}, β_j^{Del} are the log of the odd ratio of CNV j for duplication and deletion respectively.

Simulation Results

The QQ-plots of p-values of SMCKAT, MCKAT and CKAT under both simulation scenarios are presented in Figure 6.

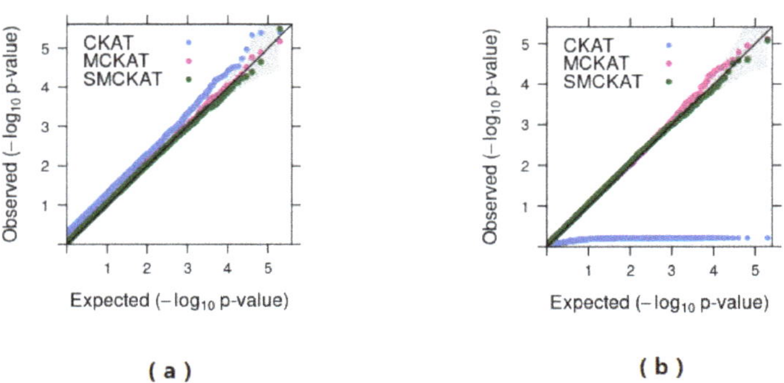

Figure 6. *p*-value based QQ-plots of MCKAT and CKAT under first (**a**) and second (**b**) simulation scenario.

Based on the QQ-plot (a), SMCKAT and MCKAT are on the 45 degree line under the first simulation scenario. This indicates that both SMCKAT and MCKAT can properly handle the type I and II error rate under different nominal significance levels even as low as 10^{-5} when dealing with the rare CNV dataset. However, CKAT is showing a higher chance of committing the type II error in detecting existing associations between the rare CNVs and phenotype.

As is shown in QQ-plot (b), both SMCKAT and MCKAT can protect the correct type I and II error rate at different nominal significance levels in dealing with the common CNV data. We observe that SMCKAT is a little conservative when the significance level is small. However, CKAT shows a weak performance in handling the type I error and detecting existing associations between the common CNVs and phenotype.

The empirical powers of SMCKAT, MCKAT and CKAT under the first and second scenarios are presented in Figures 7 and 8, respectively. As is shown is Figure 7, SMCKAT and MCKAT have almost similar powers when dealing with rare CNVs. However, CKAT shows lower power compared with SMCKAT and MCKAT. The reason might be that the CKAT is not considering the CNV dosage information when testing the association.

Similarly, in the second simulation scenario, SMCKAT and MCKAT have similar powers. However, CKAT is showing low power when dealing with common CNV data. This might be due to the CKAT scanning algorithm for aligning CNVs in the CNV profiles. The CKAT shift-by-one scanning algorithm can capture similarity between limited number of CNVs, which may result in low performance when dealing with common CNVs.

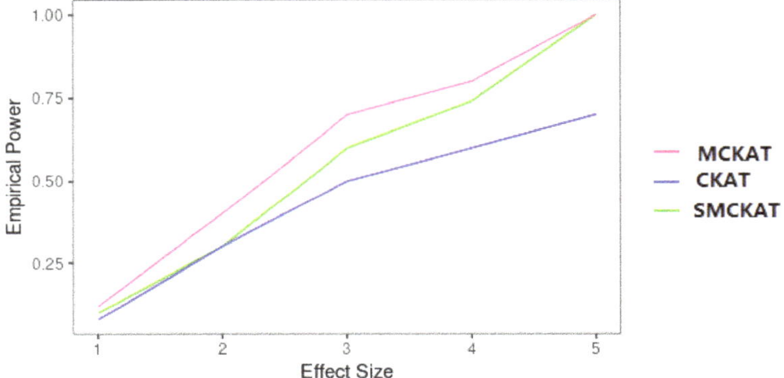

Figure 7. Empirical power of SMCKAT, MCKAT and CKAT under first simulation scenario, rare CNV data.

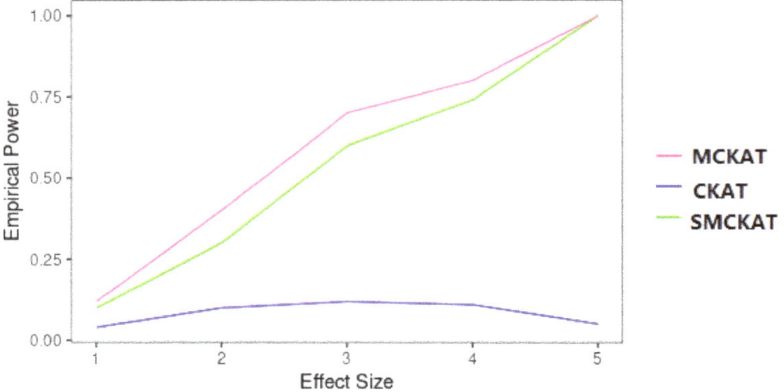

Figure 8. Empirical power of SMCKAT, MCKAT and CKAT under second simulation scenario, common CNV data.

4. Real Data Application Results

We conducted SMCKAT analysis, for different CNV group sizes, on single chromosomes and the whole genome to test the association between CNV sequential order and disease-related traits. The disease-related traits studied in this paper are cancer subtype for the RMS data set and disease status for the ASD data set. We compared SMCKAT results with those obtained from MCKAT and CKAT to evaluate SMCKAT performance on real CNV data.

4.1. CNV Analysis on Rhabdomyosarcoma Data Set

First, we conducted the experiment on the RMS data. The RMS occurs as two major histological subtypes, embryonal (ERMS) and alveolar (ARMS). The classification of the RMS subtype has a direct effect on the patient treatment options. The RMS data includes a total of 59,131 CNVs for 25 alveolar and 19 embryonal cancers. We apply SMCKAT to each of 23 chromosome pairs, with different CNV group sizes, to test the association between CNV sequential order and RMS subtype. Bonferroni correction is used for adjusting the multiple testing to control the family-wise error rate (FWER) of $\alpha = 0.05$. Since 22 chromosomes and a sex chromosome are being tested, the p-value threshold for a whole-chromosome significance is calculated as $0.05/23 = 2.2 \times 10^{-3}$. SMCKAT identifies four chromosomes out of the existing 23 chromosomes that have a CNV sequential order that is significantly associated with the RMS sub-type. The p-values of SMCKAT for these four chromosomes are reported in Table 1.

Table 1. p-values of testing the association between CNV sequential order and RMS subtype trying different CNV group sizes. n is the group size and (#) denotes the total number of CNVs on the chromosome.

Chr.	#CNV	$n = 1$	$n = 2$	$n = 3$	$n = 4$	$n = 5$	$n = 6$
2	5584	2.45×10^{-2}	5.10×10^{-2}	8.31×10^{-2}	3.49×10^{-3}	4.25×10^{-3}	3.21×10^{-2}
8	5365	2.61×10^{-5}	7.37×10^{-6}	1.13×10^{-6}	7.63×10^{-7}	4.99×10^{-8}	0
11	3449	2.03×10^{-2}	8.26×10^{-3}	2.93×10^{-3}	1.54×10^{-3}	5.82×10^{-4}	1.20×10^{-4}
13	2462	1.80×10^{-3}	3.56×10^{-3}	4.86×10^{-3}	6.06×10^{-3}	7.89×10^{-3}	6.23×10^{-2}

Based on the results, SMCKAT identifies CNV sequential order in chromosomes 2, 8, 11, and 13, significantly associated with distinguishing RMS subtype at $FWER= 2.2 \times 10^{-3}$. These results are consistent with the existing biological knowledge, which shows the ability of the SMCKAT to identify the CNV sequential order significantly associated with specific disease-related traits.

For example, ref. [14] shows that RMS is associated with specific chromosomal abnormalities that differentiate ARMS and ERMS. Based on their study, approximately 80% of ARMS tumors display a translocation between the FOXO1 transcription factor gene located on chromosome 13 and the PAX3 transcription factor gene on chromosome 2, and ERMS tumors show a higher frequency of specific genetic mutation on chromosome 11 than ARMS. Ref. [15] has revealed the same earlier. Furthermore, ref. [16] has found that the ARMS subtype is significantly associated with amplifications on chromosome 8. Our findings show another mechanism like CNVs can play a significant role in causing any disease-related traits besides gene mutations and chromosomal translocations.

We tested different CNV group sizes when applying SMCKAT to the RMS data set. Based on the results reported in Table 1, SMCKAT shows the strongest evidence and smallest p-value for the chromosome 8 for all CNV group sizes. It means subjects with the same RMS subtype may have a similar CNV sequential order on their chromosome 8.

We tested SMCKAT on the RMS data set for group sizes greater than six. We observed an increasing trend in p-values by increasing the group size, which shows a decline in the significance level of the CNV sequential order associated with the RMS subtype.

4.2. CNV Analysis on Cytogenetic Bands in RMS

Based on the result reported in Table 1, there is strong statistical evidence, as supported by a *p*-value near to zero, that the CNV sequential order of chromosome 8 with group size of six is significantly associated with the RMS subtype. Therefore, we picked chromosome 8 with a CNV group size of six for further analysis. We partitioned chromosome 8 into smaller regions based on the cytogenetic bands. We applied SMCKAT on each cytogenetic band to check if SMCKAT is capable of detecting more specific regions rather than chromosome. Then, we compared the results with of MCKAT and CKAT. Table 2 contains the *p*-value of the association test in each cytogenetic band in chromosome 8. Since 40 cytogenetic bands are being tested in chromosome 8, the *p*-value threshold for a band significance is calculated as $0.05/40 = 1.2 \times 10^{-3}$.

Table 2. *p*-values of the testing association between RMS subtype and CNVs in the chromosome 8 cytogenetic bands by SMCKAT, MCKAT and CKAT. (*) denotes significant association between RMS subtype and CNVs, (#) denotes the number of total CNVs on the band.

Arm	Band		Start	Stop	#CNVs		SMCKAT		MCKAT		CKAT
p	23	3	1	2,300,000	113	9	6×10^{-2}	3	4×10^{-4} *	4	917×10^{-1}
p	23	2	2,300,001	6,300,000	85	3	0×10^{-2}	2	0×10^{-2}	3	939×10^{-1}
p	23	1	6,300,001	12,800,000	304	1	8×10^{-4} *	4	7×10^{-8} *	4	755×10^{-1}
p	22	0	12,800,001	19,200,000	101	2	8×10^{-2}	8	2×10^{-3}	4	327×10^{-1}
p	21	3	19,200,001	23,500,000	102	1	1×10^{-1}	2	5×10^{-2}	4	237×10^{-1}
p	21	2	23,500,001	27,500,000	82	3	4×10^{-2}	3	6×10^{-2}	4	717×10^{-1}
p	21	1	27,500,001	29,000,000	50	2	5×10^{-2}	1	6×10^{-2}	4	948×10^{-1}
p	12	0	29,000,001	36,700,000	190	1	3×10^{-6} *	3	7×10^{-5} *	4	658×10^{-1}
p	11	23	36,700,001	38,500,000	48	1	0	3	7×10^{-3}	3	916×10^{-1}
p	11	22	38,500,001	39,900,000	57	9	3×10^{-2}	8	4×10^{-3}	4	$613 \times 10^{-}$
p	11	21	39,900,001	43,200,000	147	4	4×10^{-3}	1	0×10^{-4} *	3	$655 \times 10^{-}$
p	11	1	43,200,001	45,200,000	72	8	8×10^{-2}	2	8×10^{-2}	4	$584 \times 10^{-}$
q	11	1	45,200,001	47,200,000	41	1	0	2	1×10^{-2}	4	$436 \times 10^{-}$
q	11	21	47,200,001	51,300,000	200	4	4×10^{-3}	8	4×10^{-5} *	4	$064 \times 10^{-}$
q	11	22	51,300,001	51,700,000	6	9	3×10^{-1}	4	7×10^{-2}	4	$200 \times 10^{-}$
q	11	23	51,700,001	54,600,000	61	1	0	6	1×10^{-2}	4	$657 \times 10^{-}$
q	12	1	54,600,001	60,600,000	177	9	1×10^{-3}	7	0×10^{-4} *	4	$505 \times 10^{-}$
q	12	2	60,600,001	61,300,000	18	1	0	3	3×10^{-2}	4	$502 \times 10^{-}$
q	12	3	61,300,001	65,100,000	134	4	9×10^{-2}	1	1×10^{-2}	4	$110 \times 10^{-}$
q	13	1	65,100,001	67,100,000	71	4	4×10^{-2}	5	8×10^{-3}	4	$427 \times 10^{-}$
q	13	2	67,100,001	69,600,000	54	5	8×10^{-2}	4	3×10^{-3}	4	$659 \times 10^{-}$
q	13	3	69,600,001	72,000,000	62	1	4×10^{-2}	1	8×10^{-3}	3	$762 \times 10^{-}$
q	21	11	72,000,001	74,600,000	144	4	8×10^{-1}	8	4×10^{-3}	3	$325 \times 10^{-}$
q	21	12	74,600,001	74,700,000	1	1	0	1	0	1	0
q	21	13	74,700,001	83,500,000	308	1	0×10^{-2}	2	6×10^{-3}	4	$927 \times 10^{-}$
q	21	2	83,500,001	85,900,000	56	4	8×10^{-2}	2	9×10^{-2}	4	$189 \times 10^{-}$
q	21	3	85,900,001	92,300,000	185	4	7×10^{-3}	1	0×10^{-4} *	4	$215 \times 10^{-}$
q	22	1	92,300,001	97,900,000	182	1	7×10^{-2}	1	0×10^{-2}	3	$072 \times 10^{-}$
q	22	2	97,900,001	100,500,000	103	4	5×10^{-2}	3	9×10^{-3}	4	$395 \times 10^{-}$
q	22	3	100,500,001	105,100,000	162	1	2×10^{-2}	4	6×10^{-3}	4	$458 \times 10^{-}$
q	23	1	105,100,001	109,500,000	135	2	8×10^{-3}	2	5×10^{-3}	4	$017 \times 10^{-}$
q	23	2	109,500,001	111,100,000	33	9	8×10^{-1}	8	0×10^{-1}	3	$005 \times 10^{-}$
q	23	3	111,100,001	116,700,000	185	1	1×10^{-2}	2	3×10^{-3}	4	$419 \times 10^{-}$
q	24	11	116,700,001	118,300,000	53	4	6×10^{-2}	2	6×10^{-2}	4	$705 \times 10^{-}$
q	24	12	118,300,001	121,500,000	109	2	5×10^{-3}	2	2×10^{-3}	4	$068 \times 10^{-}$
q	24	13	121,500,001	126,300,000	151	2	2×10^{-2}	6	0×10^{-3}	4	$856 \times 10^{-}$
q	24	21	126,300,001	130,400,000	208	5	0×10^{-2}	1	9×10^{-2}	3	$922 \times 10^{-}$
q	24	22	130,400,001	135,400,000	155	5	5×10^{-2}	1	5×10^{-2}	4	$638 \times 10^{-}$
q	24	23	135,400,001	138,900,000	162	2	8×10^{-1}	7	7×10^{-3}	4	512×10
q	24	3	138,900,001	145,138,636	354	8	8×10^{-3}	2	5×10^{-8}*	4	277×10

As shown in Table 2, both SMCKAT and MCKAT detect significantly associated cytogenetic bands with the RMS subtype while CKAT does not identify any significant regions. MCKAT has identified 8 cytogenetic bands that CNVs on them are significantly associated with RMS sub type. Two out of these eight cytogenetic bands, 8*p*23.1 and 8*p*1 are identified by SMCKAT as well. It means not only the CNV characteristics but also the CNV sequential order in these two bands are significantly associated with the RMS sub type. Based on the results, SMCKAT has the potential to provide us with more specific CNV regions when we are testing the association between CNVs and disease-related traits compared with MCKAT.

4.3. CNV Analysis on Autism Data Set

We applied SMCKAT on the ASD data set to evaluate its performance on the rare CNV type. We aimed to test if there was any association between the sequential order of CNVs and ASD status. The ASD data set contains 1285 rare CNVs on 310 individuals with ASD and 1074 rare CNVs on 278 healthy individuals. Since the ASD data set contains only rare and large CNVs, an arbitrary CNV profile may have no or few CNVs on some chromosomes. Therefore, instead of applying SMCKAT to every 23 chromosomes, we applied it to the whole genome. Then, we tested if there is any association between the whole genome CNV sequential order and the ASD status. We considered 0.05 as the p-value threshold for the whole-chromosome significance. As is reported in Table 3, there is strong statistical evidence, up to a CNV group size of five, that subjects with the same disease status have similar CNV ordesr in their CNV profiles. We tested SMCKAT on the ASD data for the larger group sizes as well. We observed an increasing trend in p-values by increasing the group size, which shows a decline in the significance level of the CNV sequential order associated with the ASD status.

Table 3. p-values of testing the association between CNV sequential order and ASD status trying different CNV group sizes.

n	1	2	3	4	5	6
p-value	0	7.91×10^{-9}	3.09×10^{-6}	3.62×10^{-4}	4.89×10^{-3}	1.03×10^{-1}

5. Discussion

SMCKAT tests the association between CNVs and disease-related traits. It checks if CNVs are randomly distributed on the chromosomes or if their sequential orders are significant and have associations with disease-related traits. Our approach has several advantages over the existing methods. First, it measures the similarity between CNV profiles by considering not only all CNV characteristics but also the CNV sequential order. To our knowledge, it is the first approach to study the association between CNV sequential order and disease related traits. Secondly, it is applicable to both rare and common CNV data sets, while previous methods like CKAT can not deal with common CNV data sets. Thirdly, SMCKAT is more stringent when compared with the state-of-the-art approach MCKAT in detecting significant CNV regions. Finally, SMCKAT can help biologists detect significantly associated CNV regions with any disease-related trait across a patient group instead of examining the CNVs case by case in each subject.

Although our experimental results are promising and more specific compared with the state-of-the-art kernel approach, this study has limitations. There are not many publicly available CNV data sets. Besides, most available ones do not contain all CNV features together, in particular the dosage information. Consequently, our method is tested only on one data set, an RMS data set, that includes all multi-dimensional CNV characteristics. For the ASD data set, we considered a dosage less than two for all deletions and greater than two for all amplifications to make the most of the proposed method's capability. Applying SMCKAT to more data sets containing all CNV characteristics can help to determine its strengths and weaknesses. In addition, there is no existing study, neither biological nor computational, that has studied the CNV sequential order to be able to validate our experimental results with it.

Our study shows that CNV sequential order has the potential to play a significant role in causing disease-related traits, but more new findings can be revealed by conducting more comprehensive analysis upon the availability of data.

6. Conclusions

This paper presents a sequential multi-dimensional CNV association test identifying associations between CNVs and disease-rated traits using all multi-dimensional CNV characteristics and CNV sequential order. Our method, SMCKAT, uses different kernels to measure the similarity between CNV profiles with respect to both CNV orders and characteristics. Then, the similarity in CNV profiles is compared to the similarity in disease related traits to test for an association.

The evaluation was conducted on two types of CNV data sets, a rare CNV data set and a common CNV data set. Results indicate that our method provides statistically strong evidence that there is an association between the sequential order of CNVs and disease related traits. Currently, SMCKAT is capable of testing the association between CNVs and qualitative disease-rated traits. In our future work, we will expand the SMCKAT framework to be applicable to both qualitative and quantitative traits.

Author Contributions: N.M.E., P.J.K.: conceptualization and study design. N.M.E., D.C. and P.J.K data processing. N.M.E.: conducted the analysis. N.M.E.: drafting manuscript. All authors have read and agreed to the published version of the manuscript.

Funding: This research received no external funding.

Institutional Review Board Statement: All data used on this study is publilc domain data and openly accessible. Please see Data Availability Statement. Access to the RMS Data on dbGaP was provided under NCI Authorized Access #44698_6.

Informed Consent Statement: Not applicable.

Data Availability Statement: The ASD and RMS datasets supporting the conclusions of this article are available at https://www.ncbi.nlm.nih.gov/pmc/articles/PMC3213131 and https://www.ncbi.nlm.nih.gov/gap (accession number: phs000720.v3.p1) respectively (access date: 20 November 2021). The SMCKAT R package is publicly available at https://github.com/nesfehani/SMCKAT GitHub repository (access date: 20 November 2021).

Conflicts of Interest: The authors declare no conflict of interest.

References

1. National Human Genome Research Institute. *Genetics vs. Genomics Fact Sheet*; National Human Genome Research Institute: Bethesda, MD, USA, 2018. Available online: https://www.genome.gov/about-genomics/fact-sheets/Genetics-vs-Genomics (accessed on 20 November 2021).
2. Frazer, K.A.; Murray, S.S.; Schork, N.J.; Topol, E.J. Human genetic variation and its contribution to complex traits. *Nat. Rev. Genet.* **2009**, *10*, 241–251. [CrossRef]
3. Edwards, D.; Forster, J.W.; Chagné, D.; Batley, J. What Are SNPs? In *Association Mapping in Plants*; Springer: Berlin/Heidelberg, Germany, 2007; pp. 41–52.
4. Schrider, D.R.; Hahn, M.W. Gene copy-number polymorphism in nature. *Proc. R. Soc. B Biol. Sci.* **2010**, *277*, 3213–3221. [CrossRef]
5. Monlong, J.; Cossette, P.; Meloche, C.; Rouleau, G.; Girard, S.L.; Bourque, G. Human copy number variants are enriched regions of low mappability. *Nucleic Acids Res.* **2018**, *46*, 7236–7249. [CrossRef] [PubMed]
6. Zhan, X.; Girirajan, S.; Zhao, N.; Wu, M.C.; Ghosh, D. A novel copy number variants kernel association test with application autism spectrum disorders studies. *Bioinformatics* **2016**, *32*, 3603–3610. [CrossRef] [PubMed]
7. Brucker, A.; Lu, W.; West, R.M.; Yu, Q.Y.; Hsiao, C.K.; Hsiao, T.H.; Lin, C.H.; Magnusson, P.K.; Sullivan, P.F.; Szatkiewicz, J.P.; et al. Association test using Copy Number Profile Curves (CONCUR) enhances power in rare copy number variant analysis. *PLoS Comput. Biol.* **2020**, *16*, e1007797. [CrossRef]
8. Esfahani, N.M.; Catchpoole, D.; Khan, J.; Kennedy, P.J. MCKAT, a multi-dimensional copy number variant kernel association test. *BMC Bioinform.* **2021**. [CrossRef]
9. Liu, D.; Ghosh, D.; Lin, X. Estimation and testing for the effect of a genetic pathway on a disease outcome using logistic kernel machine regression via logistic mixed models. *BMC Bioinform.* **2008**, *9*, 292. [CrossRef] [PubMed]
10. Wu, M.C.; Kraft, P.; Epstein, M.P.; Taylor, D.M.; Chanock, S.J.; Hunter, D.J.; Lin, X. Powerful SNP-set analysis for case-control genome-wide association studies. *Am. J. Hum. Genet.* **2010**, *86*, 929–942. [CrossRef]

1. Davies, R.B. The distribution of a linear combination of $\chi 2$ random variables. *J. R. Stat. Soc. Ser. C Appl. Stat.* **1980**, *29*, 323–333.
2. Shern, J.F.; Chen, L.; Chmielecki, J.; Wei, J.S.; Patidar, R.; Rosenberg, M.; Ambrogio, L.; Auclair, D.; Wang, J.; Song, Y.K.; et al. Comprehensive genomic analysis of Rhabdomyosarcoma reveals a landscape of alterations affecting a common genetic axis in fusion-positive and fusion-negative tumors. *Cancer Discov.* **2014**, *4*, 216–231. [CrossRef]
3. Girirajan, S.; Brkanac, Z.; Coe, B.P.; Baker, C.; Vives, L.; Vu, T.H.; Shafer, N.; Bernier, R.; Ferrero, G.B.; Silengo, M.; et al. Relative burden of large CNVs on a range of neurodevelopmental phenotypes. *PLoS Genet.* **2011**, *7*, e1002334. [CrossRef]
4. El Demellawy, D.; McGowan-Jordan, J.; De Nanassy, J.; Chernetsova, E.; Nasr, A. Update on molecular findings in rhabdomyosarcoma. *Pathology* **2017**, *49*, 238–246. [CrossRef]
5. Sun, X.; Guo, W.; Shen, J.K.; Mankin, H.J.; Hornicek, F.J.; Duan, Z. Rhabdomyosarcoma: Advances in molecular and cellular biology. *Sarcoma* **2015**, *2015*, 232010. [CrossRef]
6. Nishimura, R.; Takita, J.; Sato-Otsubo, A.; Kato, M.; Koh, K.; Hanada, R.; Tanaka, Y.; Kato, K.; Maeda, D.; Fukayama, M.; et al. Characterization of genetic lesions in Rhabdomyosarcoma using a high-density single nucleotide polymorphism array. *Cancer Sci.* **2013**, *104*, 856–864. [CrossRef]

Article

Genome-Wide Scanning of Potential Hotspots for Adenosine Methylation: A Potential Path to Neuronal Development

Sanjay Kumar [1,†], Lung-Wen Tsai [2,3,4,†], Pavan Kumar [5,6], Rajni Dubey [2], Deepika Gupta [7], Anjani Kumar Singh [8], Vishnu Swarup [7,*] and Himanshu Narayan Singh [9,*]

1 Department of Life Sciences, School of Basic Sciences and Research, Sharda University, Greater Noida 201310, India; Sanjay.Kumar7@sharda.ac.in
2 Department of Medicine Research, Taipei Medical University Hospital, Taipei 11031, Taiwan; lungwen@tmu.edu.tw (L.-W.T.); 205095@h.tmu.edu.tw (R.D.)
3 Department of Information Technology Office, Taipei Medical University Hospital, Taipei 11031, Taiwan
4 Graduate Institute of Data Science, College of Management, Taipei Medical University, Taipei 11031, Taiwan
5 Department of Anatomy, All India Institute of Medical Sciences, New Delhi 110029, India; kumarpa@uic.edu
6 Department of Anatomy & Cell Biology, College of Medicine, University of Illinois, Chicago, IL 60612, USA
7 Department of Neurology, All India Institute of Medical Sciences, New Delhi 110029, India; deepa12aug@gmail.com
8 Department of Physics, Atma Ram Sanatan Dharma College, University of Delhi, New Delhi 110021, India; aksingh@arsd.du.ac.in
9 Department of System Biology, Columbia University Irving Medical Center, New York, NY 10032, USA
* Correspondence: vishnuswarup@gmail.com (V.S.); hs3290@columbia.edu (H.N.S.)
† These authors contributed equally.

Citation: Kumar, S.; Tsai, L.-W.; Kumar, P.; Dubey, R.; Gupta, D.; Singh, A.K.; Swarup, V.; Singh, H.N. Genome-Wide Scanning of Potential Hotspots for Adenosine Methylation: A Potential Path to Neuronal Development. *Life* **2021**, *11*, 1185. https://doi.org/10.3390/life11111185

Academic Editors: Md. Altaf-Ul-Amin, Shigehiko Kanaya, Naoaki Ono and Ming Huang

Received: 26 September 2021
Accepted: 30 October 2021
Published: 5 November 2021

Abstract: Methylation of adenosines at N6 position (m6A) is the most frequent internal modification in mRNAs of the human genome and attributable to diverse roles in physiological development, and pathophysiological processes. However, studies on the role of m6A in neuronal development are sparse and not well-documented. The m6A detection remains challenging due to its inconsistent pattern and less sensitivity by the current detection techniques. Therefore, we applied a sliding window technique to identify the consensus site (5′-GGACT-3′) $n \geq 2$ and annotated all m6A hotspots in the human genome. Over 6.78×10^7 hotspots were identified and 96.4% were found to be located in the non-coding regions, suggesting that methylation occurs before splicing. Several genes, *RPS6K*, *NRP1*, *NRXN*, *EGFR*, *YTHDF2*, have been involved in various stages of neuron development and their functioning. However, the contribution of m6A in these genes needs further validation in the experimental model. Thus, the present study elaborates the location of m6A in the human genome and its function in neuron physiology.

Keywords: adenosine methylation; m6A; RNA modification; neuronal development

1. Introduction

Among the 150 reported RNA modifications to date, methylation at N6 position of adenosine (m6A) is the post-transcriptional RNA modification with a high physiological relevance [1]. This reversible modification of RNA regulates the expression of several genes and affects human physiology [2]. Over 7000 genes have been reported to carry this modification in humans, and aberrant RNA modification contributes to the pathogenesis of various human diseases. Notably, the abnormal modification of human tRNA may lead to mental retardation and intellectual disability [3]. Among all different RNA modifications, m6A modification is most abundant in mRNAs of eukaryotic cells. Altered m6A modifications have been linked with several diseases, such as obesity, cancer, diabetes mellitus, stress-related psychiatric disorders, neuronal development, and functions [4,5]. Several analytical tools have revealed that 5′-GGACU-3′ is the most common structural signature for m6A modification [6,7].

Recent reports demonstrate that not all the adenines in RNA are methylated; the probability of methylation is random, and some RNAs are even entirely devoid of this modification. Moreover, no consensus has been reached for the methylation pattern; nucleotides flanking to "methylable adenines" impact the possibility of their methylation. Cumulatively, these factors cause difficulties in the analysis during in vitro validation of m6A in RNA. In addition, there are several limitations in the current technologies, which are being used for identification of m6A sites. The resolution of methyl-RNA immune-precipitation and sequencing (MeRIP-Seq) covers around 200 nucleotides; therefore, it cannot be used to pinpoint the precise location of the m6A modification [8]. Another technique called site-specific cleavage and radioactive-labeling followed by ligation-assisted extraction and thin-layer chromatography (SCARLET) is time-consuming and expensive and not feasible for high-throughput applications [9,10]. Most existing methods are entirely ineffective in identifying m6A sites due to a biassing and unpredictability of chemicals toward a specific RNA modification, and failure to produce single-nucleotide sequencing data [11–13]. Intrinsic features, such as fragility, multiple open reading frames, alternative splicing, and short RNA half-lives contribute to these m6A analysis flaws. Thus, generating all potential m6A sites in a single transcriptome analysis within a predefined time frame is challenging with these currently available tools. Alternatively, tagging the target sequence in the genome itself can unveil the distribution of all potential m6A sites, which display methylation possibilities, and perhaps aiding in the understanding of m6A's function in physiological processes. Here, we present the sliding window-based technique to identify all adenines in the human genome, considering each one as a potential methylation site. Furthermore, we have also delineated the role of m6A modification in the neurological milieu, contrasting the physiological and pathological conditions.

2. Methodology

2.1. Definition of m6A Methylation Sites

The consensus sequence (5′-GGACT-3′)n, n = 2 in tandem was searched throughout the human genome (version GRCh37 patch 8). If methylated, the two consensus sequence in tandem are considered as more effective in generating physiological effects. Following the strict criteria, no mismatch in the m6A sites was allowed.

2.2. PatternRepeatAnnotator: A Home-Made PERL Script

To locate m6A sites in the human genome, a home made PERL script, named "PatternRepeatAnnotator" based on the sliding window technique or window shift algorithm was used [14,15]. The "PatternRepeatAnnotator" was developed to explore the user-defined patterns in the genome sequence (Figure 1). The sliding window technique is a method for finding a subarray (e.g., consensus sequence) in the genome that satisfies the given conditions (e.g., tandem). The search was carried out by maintaining a subset of items (e.g., nucleotides) as a window, and rearranged accordingly and shifted them within the more extensive list until the subarray is precisely matched. The "PatternRepeatAnnotator" scanned the consensus sequences through each chromosome (in Fasta format) to locate them with a particular length (n) defined by the user. Consequently, it provided chromosome-wise coordinates for all the identified sites.

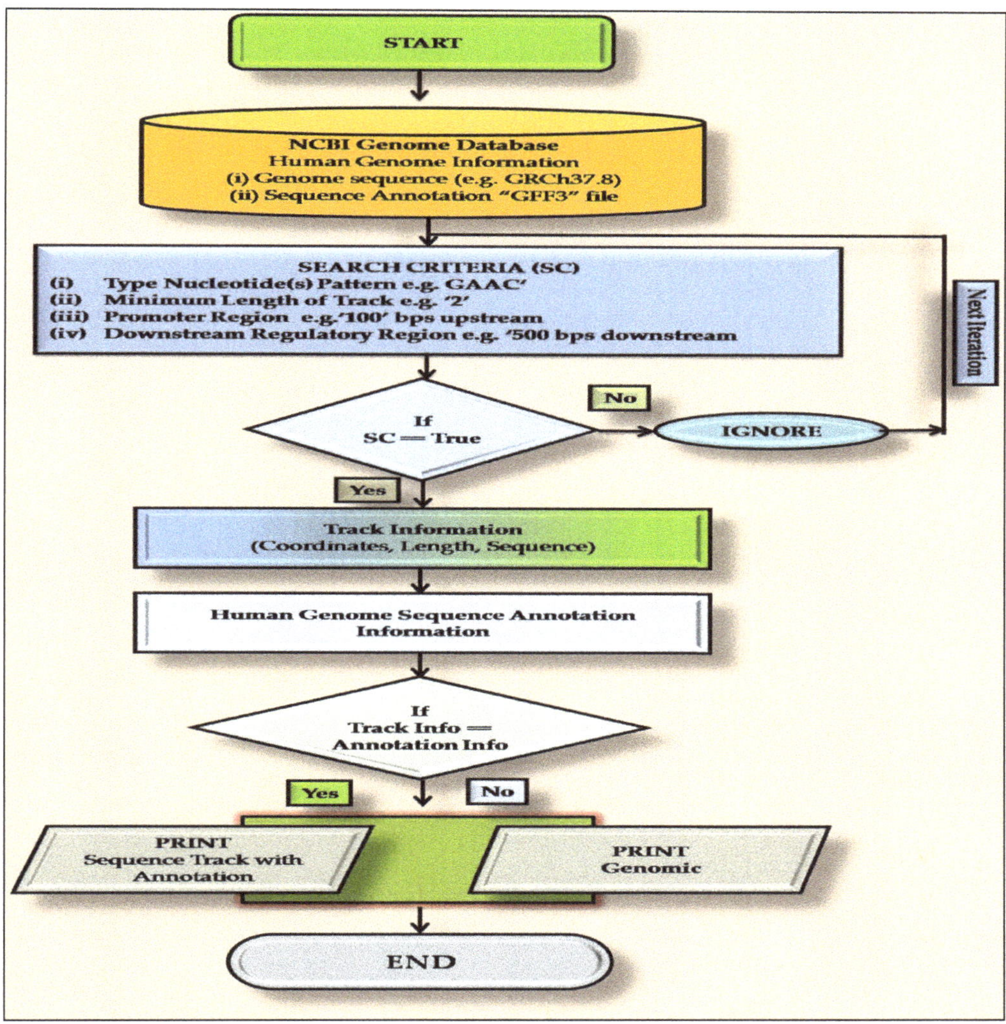

Figure 1. Schematic algorithm used to develop the "PatternRepeatAnnotator".

2.3. Annotation of m6A Sites

To annotate the identified m6A sites, the GRCh37 genome annotation file was utilized (https://ftp.ncbi.nlm.nih.gov/genomes/archive/old_refseq/Homo_sapiens/ARCHIVE/BUILD.37.3/GFF/ref_GRCh37.p5_top_level.gff3.gz, accessed on 26 September 2021). The identified coordinates of m6A sites were further mapped to the annotation file. After the processing, all information was transported to a comma-separated value (.csv) file, where the running task was conducted. The promoter and downstream regulatory regions (DRR) were considered as 100 nucleotides upstream and 500 nucleotides downstream of all identified genes, respectively. The genes containing recognition sequences in the coding (plus/sense) DNA strand were selected for further analysis only. A single gene was counted as one entry, even if it had the target sequence at multiple locations.

2.4. Gene Ontology (GO) Analysis

To assess the mechanistic biological insight into the genes of interest, Gene Ontology (GO) analysis was performed using gprofiler [16]. Enrichment maps were generated using ShinyGo, a gene ontology enrichment analysis software (South Dakota State University Bioinformatics Research group). The distribution of target sequences ($n \geq 2$) in protein coding genes with their frequencies and enrichment score per Mb of respective chromosome were analyzed.

3. Results

A total of 6.78×10^7 target sequences GGACT ($n \geq 2$) were found throughout the human genome using the homemade script. Chromosome 2, having 242 million base pairs (Mbps) nucleotides were found to carry the highest number of target sequences in total ($n = 1014.79 \times 10^4$). Out of these, the target sequences of 31.76×10^4, 541.56×10^4, 1.45×10^4, 433.77×10^4,and 6.23×10^4 Mbps were found in exonic, intronic, promoter, genomic, and downstream regulatory regions (DRR), respectively (Table 1, Figure 2a). The enrichment (copy number of target sequence per Mbps of the chromosome) of target sequence was also found to be highest (4.19×10^4 sequences/Mbps) in chromosome 2 (Figure 2b). Chromosome 24 was found to carry the lowest number of target sequences in total 41.2×10^4 Mbps with an enrichment score of 0.72×10^4. Out of these, the target sequences 0.07×10^4, 0.31×10^4, 0.67×10^4, 10.31×10^4, and 29.93×10^4 Mbps were identified in promoter, DRR, exonic, intronic, and genomic regions, respectively (Table 1).

Table 1. Distribution of target sequence ($n \geq 2$) found in different regions of human genome.

Chromosome Number	Number of Target Sequence $\times 10^4$					
	Promoter	DRR	Exon	Intron	Genomic	Total
1	1.00	4.17	22.08	289.36	202.29	518.90
2	1.46	6.23	31.76	541.57	433.78	1014.80
3	0.51	2.13	11.55	229.46	142.93	386.58
4	0.90	3.92	18.34	368.27	391.23	782.67
5	0.14	0.13	2.95	60.49	79.17	142.89
6	0.63	0.54	11.49	131.76	108.23	252.65
7	0.38	0.33	7.74	127.44	108.97	244.86
8	0.32	0.27	6.31	103.02	79.42	189.34
9	0.11	0.10	2.29	56.21	50.51	109.22
10	0.23	0.20	4.89	91.10	69.49	165.92
11	1.16	4.89	23.13	293.85	238.57	561.61
12	0.27	0.23	5.90	82.65	55.61	144.66
13	0.52	0.45	9.64	183.52	205.59	399.72
14	0.80	0.68	13.88	194.32	168.73	378.41
15	0.71	0.59	15.63	208.53	129.65	355.11
16	0.42	0.32	8.76	88.48	59.10	157.08
17	0.30	0.24	6.60	57.25	34.28	98.67
18	0.10	0.09	2.06	34.53	27.32	64.10
19	0.44	0.37	9.38	61.79	37.57	109.54
20	0.19	0.16	3.66	56.90	50.03	110.93
21	0.24	0.21	4.74	64.69	79.52	149.41
22	0.47	0.39	9.70	93.20	54.50	158.26
23	0.31	0.28	6.19	105.08	135.69	247.54
24	0.07	0.31	0.67	10.31	29.93	41.29
Total	11.68	27.23	239.34	3533.78	2972.11	6784.1
Percentage of Total	0.172	0.401	3.528	52.089	43.810	100.00

DRR—Downstream Regulatory Regions.

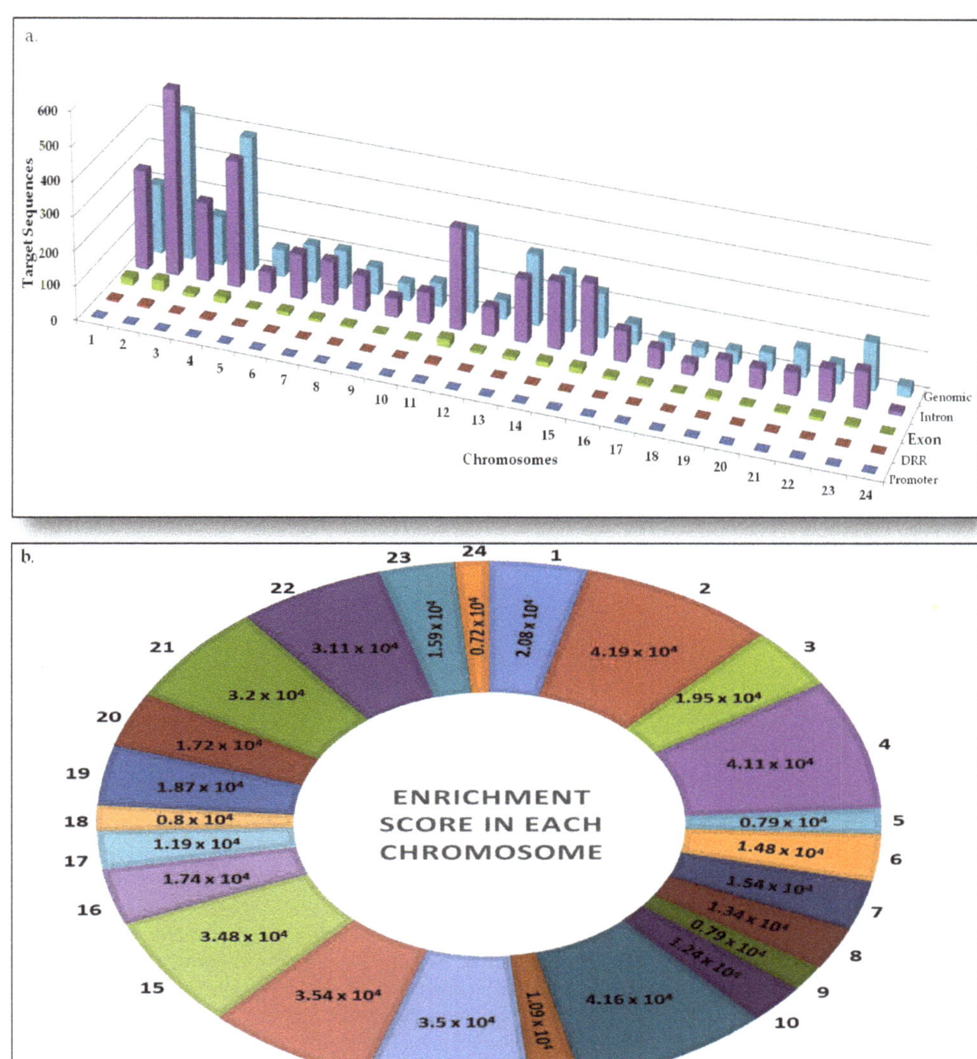

Figure 2. Distribution and enrichment score of m6A sites (**a**). The potential m6A sites ($\times 10^4$) in different parts of human genome, such as promoters, DRR, exons, and genomic (intergenic) regions. (**b**) Enrichment score of target sequences according to chromosome size (in million bases pair). DRR: Downstream regulatory regions.

Subsequently, we also looked up the protein-coding genes per chromosome, which carry the target sequence ($n \geq 2$). Here, chromosome 2 had the highest number of genes ($n = 1448$) with the target sequence followed by chromosome 11 ($n = 982$) (Table 2). Interestingly, a notable highest frequency of the target sequence ($n = 163$) was observed in MCF2 Transforming Sequence-Like (*MCF2L*) gene located on chromosome 13. Additionally, the highest number of protein-coding genes were also found on chromosome 13 (81%; 266/327), followed by chromosome 4 (76%; 572/752), whilst chromosome 9 had the lowest number of protein-coding genes with the target sequence (8%; 64/786). Notably, the chromosome 1, containing the highest number of protein-coding genes (n = 2058), was found to carry the target sequence only in 27% of genes (Table 2).

Table 2. Distribution of target sequences ($n \geq 2$) in protein-coding genes with their frequencies and enrichment score per Mb of respective chromosomes.

Chromosome	Chromosome Size (Mb)	Total No. Protein Coding Genes Present	Number of Protein Coding Genes Carrying Target Sequence (%)	Highest Frequency of Target Sequence in Any Gene	# Enrichment Score $\times 10^4$
1	249	2058	967 (27)	63	2.08
2	242	1309	1448 (67)	58	4.19
3	198	1078	522 (30)	62	1.95
4	190	752	932 (76)	55	4.11
5	182	876	135 (10)	64	0.79
6	171	1048	497 (26)	32	1.48
7	159	989	352 (21)	51	1.54
8	145	677	286 (25)	73	1.30
9	138	786	99 (8)	88	0.79
10	134	733	226 (18)	43	1.24
11	135	1298	982 (42)	73	4.16
12	133	1034	265 (14)	36	1.09
13	114	327	432 (81)	163	3.50
14	107	830	587 (40)	74	3.54
15	102	613	641 (64)	40	3.48
16	90	873	343 (19)	108	1.74
17	83	1197	261 (12)	21	1.19
18	80	270	92 (18)	35	0.80
19	59	1472	361 (13)	12	1.87
20	64	544	169 (20)	69	1.72
21	47	234	212 (56)	47	3.20
22	51	488	39 (44)	34	3.11
23	156	842	238 (17)	80	1.59
24	57	71	42 (24)	14	0.72

Enrichment score was calculated as copy number of target sequence per Mbps of chromosome.

Here, the consensus site (5′-GGACT-3′) $n \geq 2$ was utilized to locate and annotate a m6A hotspots. We identified several genes associated to cancer, diabetes, stress-relate mental illnesses, and neuronal development, among other diseases. Especially, GO analys revealed the crucial genes related to neuronal development.

m6A RNA modification is one of the most prevalent reversible internal modificatior regulated by methyltransferases ("writers") and demethylases ("erasers") [17]. The pre ence of complementary seed sequences in micro-RNAs (miRNAs) indicated that miRN/ targeted m6A peak regions in both mouse and human experimental studies.Furthermor m6A has also been reported in the transcriptome of neurons [9,18]. Brain development is highly specific and coordinated genetic event andany abnormalities can act as a doorw to different anomalies, such as autistic spectrum and schizophrenia-like disorders [19–2 In our GO analysis data, we selected 1729 genesbased on frequency of target sequen (GGACT) more than 2.Of them, only 27 were scrutinized. The enrichment analysis of t biological process for m6A hotspot genes revealedits association with embryonic bra development, locomotion, neuronal projection, neuronal differentiation, axonal guidan synaptic assembly, synaptic plasticity, and transmission (Figure 3a,b).

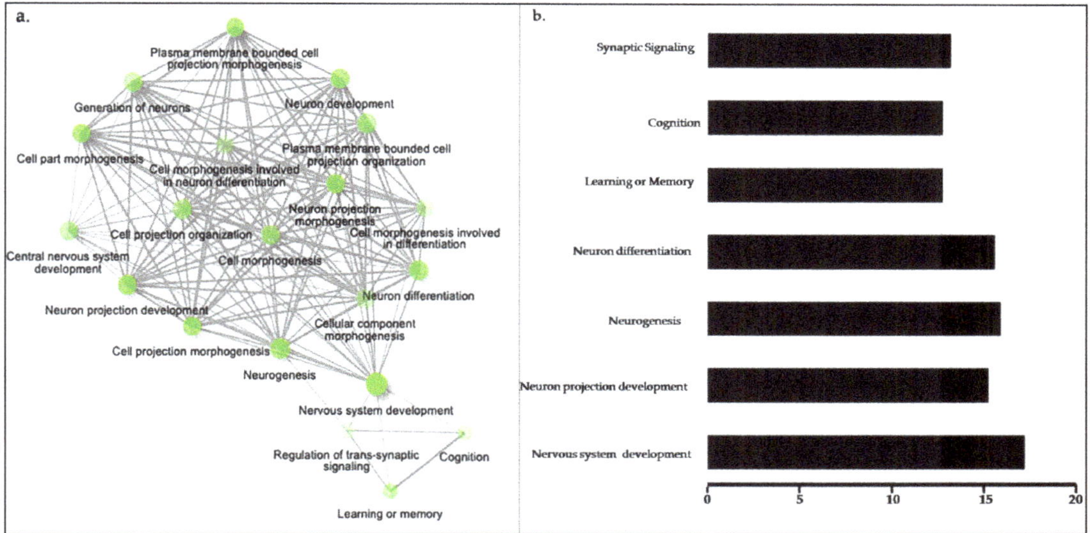

Figure 3. GO analysis of m6A target sites. (**a**) The networking analysis of m6A hotspot genes in different physiological processes. (**b**) Enrichment analysis of m6A hotspot genes for neurological processes, such as neuronal development, neurogenesis, differentiation and projection.

4. Discussion

The human genome sequence was explored for all possible m6A sites with two or more target sequences (5′-GGACT-3′) in tandem, which might have a high probability for methylation. The human genome may include some m6A-containing motifs, that still remain unidentified due to their less abundance or beyond the range of advanced detection techniques; hence, surveying the human genome for target sites could be an alternative tool to identify them.

Using the tool "PatternRepeatAnnotator", a total of 6.78×10^7 target sequences were recognized on the plus strand of the human genome. We observed over representation of the target sequences in non-coding DNA (96.4% in introns, DRR, promoters and genomic regions), whereas a small quantity of 3.5% was located in coding (exonic) regions (Supplementary Figure S1). This internal modification has been reported in nascent pre-mRNAs, suggesting that the addition of methylation group occurs before splicing [22], which is supported by our current findings with 52% target sequences in intronic regions. The m6A modification exhibits spatio-temporal specific expression patterns; therefore, despite many target sequences, only a few undergo methylation [23]. The high density of m6A sites present in 95.8% of intron in non-coding genomic regions, were primarily involved in producing miRNAs. It has been reported that miRNAs influence the fundamental biological processes from cell division to cell death and may undergo m6A modification [24]. For example, m6A modifications in primary miRNA enhance their recognition and processing by DGCR8, a miRNA microprocessor complex protein [25]. Therefore, identified m6A sites may provide deep insight into the mRNA–miRNA interaction pathways involved in the pathogenesis of various diseases. Ribosomal protein S6 kinase genes *RPS6K* have been predicted as a potential candidate for the pathogenesis of hepatocellular carcinoma by the miRNA–mRNA network analysis [26]. This is in line with our enrichment analysis (Supplementary Table S1) identifying RPS6KA3 and RPS6KA5 ribosomal genes, which are associated with regulation of axonogenesis and cellular morphogenesis in the course of neuronal differentiation. Any alteration of m6A methylation of *RPS6KA3* and *RPS6KA5* may affect the normal neurite outgrowth and arborization [27].

Neurexin performs distinct regulatory functions in different classes of neurons, and any mutation or deletion of Neurexin (*NRXN1* and *NRXN2*) genes have been associated with autism-associated behavioral changes in experimental mice [28]. Neurexin also plays a key role in the trafficking of presynaptic vesicles and their deletion resulted in the reduction of synaptic current. To our knowledge, no report exists on the direct link between neurexins and m6A. However, our enrichment analysis data have shown that m6A may regulate *NRXN1*, *NRXN2* and *NRXN3* genes.

In a synaptic epi-transcriptomic study, 4469 enriched m6A sites have been reported selectively in 2921 genes in the forebrain of adult mice and imply that chemically modified mRNA could significantly promote synaptic function [29]. The knockdown of the m6A reader has shown a dramatic change in the spine morphology and dampened the synaptic transmission, there by suggesting its role in synaptic function. Epidermal Growth Factor Receptor (EGFR) belongs to the tyrosine kinase family and is expressed by neuronal and glial cells in different brain regions [30]. During the early development, EGFR is highly expressed in the midbrain and hippocampus, and its increased expression has been also reported in many pathophysiologies, including Alzheimer's, Huntington's, Parkinson disease, amyotrophic lateral sclerosis, and traumatic brain injury associated with reactive gliosis [31]. Our data have also shown that m6A is enriched with EGFR, which is consistent with previous findings [32]. YT521-B homology domain family 2 (*YTHDF2*) is a m6A reader and directly binds the m6A modification site of EGFR 3′UTR of mRNA and impedes cell proliferation and growth by modulating the downstream ERK/MAPK pathway [32]. The functions of EGFR could also be modulated by other proteins such as *METTL3* and *FTO* [33,34]. Collectively, these data indicated that m6A modification of mRNA is requisite for the proper physiological functions of EGFR. Further, the MAPK is a key regulator of neurogenesis, which consists of four distinct cascades, ERK1/2, JNK1/2/ p38, and ERK5. It has been shown that m6A enriched with *MAPK* and *METTL* played a tumour-suppressive role via the p38/ERK pathway. Since, elevated levels of p-38 and pERK in colorectal cancer have displayed the inhibition of cell migration and proliferation after knockdown of METTL [35]. Likewise, *EGFR, YTHDF2* also regulate the MAPK and NF-kB signalling in systemic lupus erythematosus (SLE). YTHDF2 knockdown has been demonstrated to activate MAPK and NF-kB and resulted in a significant increase in pro-inflammatory events in SLE [7,36]. Additionally, the neurological involvement appears the early stage in SLE, with cognitive impairment being the most prevalent symptom that correlates with disease activity [37].

The identification and quantification of m6A in the transcriptome are tedious, expensive, and associated with many significant systematic errors. To date, well established in vitro methods have encountered several obstacles, including single-nucleotide resolution, a lack of selective chemical reactivities for a specific RNA modification, and lengthy protocols for m6A identification. These challenges are exacerbated by the stability RNA and the random frequency of methylation. As a result, finding m6A signatures throughout the whole transcriptome is an extremely difficult task. To address these issues, several webtools and algorithms have been developed, which either investigate various databases of m6A sequences or utilize statistical techniques to more precisely locate m6A sites [36,38–42]. Other tools, such as iRNA-AI, iMethyl-PseAAC, iDNA-Methyl, iRNA-Methyl, and iRNA-PseU have been generated also for the identification and annotation specific sites for adenosine to inosine editing, protein methylation, DNA methylation, N6 methyl adenosine, using pseudo-nucleotide, and RNA pseudouridine, respectively [42–46]. These tools need a sequence of interest in which the intended modification is sought, and they offer information on whether or not the desired change is feasible in that sequence. The method created in this work scanned the whole human genome for identification of specific set of nucleotides (target sequence) and generated well-annotated information as output. This tool fundamentally differs in the origin of the hypothesis, concept of algorithm, and the final results compared with all other available techniques.

The Perl-script-based tool "PatternRepeatAnnotator" employed in our study can be customized in several ways: (i) it can be used to search any repeat type (e.g., CAG triplet repeats of Huntington's disease, GAA repeats of Friedreich's ataxia, etc.), (ii) the number of such repeats (1 or more) in tandem can be chosen by the user, (iii) range of promoter/downstream regions (in nucleotide length) can be given at user's choice, (iv) more importantly, the tool is futuristic, and the latest human genome version (>GRCh37 patch 8) can be provided as a template for target sequence search. The results are stored in a specified folder name after the input sequence, where numerous statistical tools can be applied to analyze data easily. The output file contains well-annotated information, such as (i) identified target sequence viz gene ID, (ii) its symbol, (iii) strand (plus/minus), (iv) location in chromosome (exon/intron/genomic/promoter/downstreamregions), (v) the position of repeat (start to end), (vi) its total length (nucleotides long) and (vi) the sequence itself. Using this robust annotated information, the analysis becomes easier, and the genes of interest can be directly picked up from the desired chromosome for further analysis. This, in turn, reduces the cost, time, and manpower required to evaluate the whole transcriptome for m6A modification. The ability to analyze databases in future depicts long-lived applicability, highly customizable interface, making it user-friendly and robust with rich annotated data.

5. Conclusions

The m6A is a conservative phenomenon and has been involved in modulating translation efficiency, mRNA turnover, RNA splicing, miRNA and other non-coding RNA biogenesis. As demonstrated in our study, "PatternRepeatAnnotator" could identify and annotate all "methylable adenosines" in the genome, however, their regulation in vivo needs to be verified as not all m6A sites are modified in the human genome. Annotation of these identified m6A sites revealed that over 96% m6A were found in non-coding regions, which corroborates their roles in downstream regulatory processes. Several essential genes in neuronal development harbor extensive m6A sites. More in vivo investigations are required to correlate these identified m6A sites, their modification pattern, and mechanistic approach in cellular processes and various human diseases.

Supplementary Materials: The following are available online at https://www.mdpi.com/article/10.3390/life11111185/s1, Figure S1: Percentage distribution of target sequences in different regions of human genome. Table S1: Enrichment Analysis of genes for their biological functions.

Author Contributions: Conceptualization, S.K. and H.N.S.; data curation, L.-W.T., D.G., V.S. and H.N.S.; resources, A.K.S.; supervision, V.S. and H.N.S.; validation, S.K., L.-W.T., D.G., R.D., V.S. and H.N.S.; visualization, S.K., R.D.; writing—original draft, P.K.; writing—review and editing, S.K., L.-W.T., R.D., D.G., V.S. and H.N.S. All authors have read and agreed to the published version of the manuscript.

Funding: None.

Institutional Review Board Statement: This article does not contain any studies involving human or animal participants.

Informed Consent Statement: This article does not contain any studies involving human or animal participants. Therefore, this is not required.

Data Availability Statement: Not applicable.

Acknowledgments: All authors aknowledge the Sharda University-UP, AIIMS-New Delhi and MTA infotech-Varanasi for providing all resources required for this study.

Conflicts of Interest: The authors declare that there are no conflict of interest.

References

Hussain, S.; Aleksic, J.; Blanco, S.; Dietmann, S.; Frye, M. Characterizing 5-Methylcytosine in the Mammalian Epitranscriptome. *Genome Biol.* **2013**, *14*, 215. [CrossRef]

Jia, G.; Fu, Y.; He, C. Reversible RNA Adenosine Methylation in Biological Regulation. *Trends Genet.* **2013**, *29*, 108–115. [CrossRef] [PubMed]

3. Bednářová, A.; Hanna, M.; Durham, I.; VanCleave, T.; England, A.; Chaudhuri, A.; Krishnan, N. Lost in Translation: Defects in Transfer RNA Modifications and Neurological Disorders. *Front. Mol. Neurosci.* **2017**, *10*, 135. [CrossRef] [PubMed]
4. Wei, W.; Ji, X.; Guo, X.; Ji, S. Regulatory Role of N6-Methyladenosine (M6A) Methylation in RNA Processing and Human Diseases. *J. Cell. Biochem.* **2017**, *118*, 2534–2543. [CrossRef]
5. Min, K.; Zealy, R.W.; Davila, S.; Fomin, M.; Cummings, J.C.; Makowsky, D.; Mcdowell, C.H.; Thigpen, H.; Hafner, M.; Kwon, S.; et al. Profiling of M6A RNA Modifications Identified an Age-associated Regulation of AGO2 MRNA Stability. *Aging Cell* **2018**, *17*, e12753. [CrossRef]
6. Spitale, R.C.; Flynn, R.A.; Zhang, Q.C.; Crisalli, P.; Lee, B.; Jung, J.-W.; Kuchelmeister, H.Y.; Batista, P.J.; Torre, E.A.; Kool, E.T.; et al. Structural Imprints in Vivo Decode RNA Regulatory Mechanisms. *Nature* **2015**, *519*, 486–490. [CrossRef]
7. Liu, J.; Yue, Y.; Han, D.; Wang, X.; Fu, Y.; Zhang, L.; Jia, G.; Yu, M.; Lu, Z.; Deng, X.; et al. A METTL3-METTL14 Complex Mediates Mammalian Nuclear RNA N6-Adenosine Methylation. *Nat. Chem. Biol.* **2014**, *10*, 93–95. [CrossRef] [PubMed]
8. Linder, B.; Grozhik, A.V.; Olarerin-George, A.O.; Meydan, C.; Mason, C.E.; Jaffrey, S.R. Single-Nucleotide-Resolution Mapping of M6A and M6Am throughout the Transcriptome. *Nat. Methods* **2015**, *12*, 767–772. [CrossRef]
9. Chen, K.; Lu, Z.; Wang, X.; Fu, Y.; Luo, G.-Z.; Liu, N.; Han, D.; Dominissini, D.; Dai, Q.; Pan, T.; et al. High-Resolution N(6)-Methyladenosine (m(6) A) Map Using Photo-Crosslinking-Assisted m(6) A Sequencing. *Angew. Chem. Int. Ed. Engl.* **2015**, *54*, 1587–1590. [CrossRef]
10. Hengesbach, M.; Meusburger, M.; Lyko, F.; Helm, M. Use of DNAzymes for Site-Specific Analysis of Ribonucleotide Modifications. *RNA* **2008**, *14*, 180–187. [CrossRef]
11. Novoa, E.M.; Mason, C.E.; Mattick, J.S. Charting the Unknown Epitranscriptome. *Nat. Rev. Mol. Cell Biol.* **2017**, *18*, 339–340. [CrossRef] [PubMed]
12. Jonkhout, N.; Tran, J.; Smith, M.A.; Schonrock, N.; Mattick, J.S.; Novoa, E.M. The RNA Modification Landscape in Human Disease. *RNA* **2017**, *23*, 1754–1769. [CrossRef] [PubMed]
13. Delatte, B.; Wang, F.; Ngoc, L.V.; Collignon, E.; Bonvin, E.; Deplus, R.; Calonne, E.; Hassabi, B.; Putmans, P.; Awe, S.; et al. Transcriptome-Wide Distribution and Function of RNA Hydroxymethylcytosine. *Science* **2016**, *351*, 282–285. [CrossRef]
14. Tanbeer, S.K.; Ahmed, C.F.; Jeong, B.-S.; Lee, Y.-K. Sliding Window-Based Frequent Pattern Mining over Data Streams. *Inf. Sc* **2009**, *179*, 3843–3865. [CrossRef]
15. Singh, H.N.; Rajeswari, M.R. NTrackAnnotator: Software for Detection and Annotation of Sequence Tracks of Chosen Nucleic Acid Bases with Defined Length in Genome. *Gene Rep.* **2017**, *7*, 32–34. [CrossRef]
16. Ge, S.X.; Jung, D.; Yao, R. ShinyGO: A Graphical Gene-Set Enrichment Tool for Animals and Plants. *Bioinformatics* **2020**, *3* 2628–2629. [CrossRef]
17. Jiang, X.; Liu, B.; Nie, Z.; Duan, L.; Xiong, Q.; Jin, Z.; Yang, C.; Chen, Y. The Role of M6A Modification in the Biological Function and Diseases. *Sig. Transduct. Target.* **2021**, *6*, 74. [CrossRef]
18. Livneh, I.; Moshitch-Moshkovitz, S.; Amariglio, N.; Rechavi, G.; Dominissini, D. The M6A Epitranscriptome: Transcriptome Plasticity in Brain Development and Function. *Nat. Rev. Neurosci.* **2020**, *21*, 36–51. [CrossRef]
19. Okano, H.; Temple, S. Cell Types to Order: Temporal Specification of CNS Stem Cells. *Curr. Opin. Neurobiol.* **2009**, *1* 112–119. [CrossRef]
20. Ohi, K.; Shimada, T.; Nitta, Y.; Kihara, H.; Okubo, H.; Uehara, T.; Kawasaki, Y. Specific Gene Expression Patterns of 1C Schizophrenia-Associated Loci in Cortex. *Schizophr. Res.* **2016**, *174*, 35–38. [CrossRef]
21. Yoon, K.-J.; Ringeling, F.R.; Vissers, C.; Jacob, F.; Pokrass, M.; Jimenez-Cyrus, D.; Su, Y.; Kim, N.-S.; Zhu, Y.; Zheng, L.; et al. Temporal Control of Mammalian Cortical Neurogenesis by M6A Methylation. *Cell* **2017**, *171*, 877–889.e17. [CrossRef] [PubMed]
22. Ke, S.; Pandya-Jones, A.; Saito, Y.; Fak, J.J.; Vågbø, C.B.; Geula, S.; Hanna, J.H.; Black, D.L.; Darnell, J.E.; Darnell, R.B. M6A MRNA Modifications Are Deposited in Nascent Pre-MRNA and Are Not Required for Splicing but Do Specify Cytoplasmic Turnov *Genes Dev.* **2017**, *31*, 990–1006. [CrossRef]
23. Meyer, K.D.; Saletore, Y.; Zumbo, P.; Elemento, O.; Mason, C.E.; Jaffrey, S.R. Comprehensive Analysis of MRNA Methylation Reveals Enrichment in 3' UTRs and near Stop Codons. *Cell* **2012**, *149*, 1635–1646. [CrossRef] [PubMed]
24. Berulava, T.; Rahmann, S.; Rademacher, K.; Klein-Hitpass, L.; Horsthemke, B. N6-Adenosine Methylation in MiRNAs. *PLoS ON* **2015**, *10*, e0118438. [CrossRef] [PubMed]
25. Alarcón, C.R.; Lee, H.; Goodarzi, H.; Halberg, N.; Tavazoie, S.F. N6-Methyladenosine Marks Primary MicroRNAs for Processin *Nature* **2015**, *519*, 482–485. [CrossRef] [PubMed]
26. Wang, W.; Zhao, L.J.; Tan, Y.-X.; Ren, H.; Qi, Z.-T. Identification of Deregulated MiRNAs and Their Targets in Hepatitis Virus-Associated Hepatocellular Carcinoma. *World J. Gastroenterol.* **2012**, *18*, 5442–5453. [CrossRef]
27. Su, L.; Song, X.; Xue, Z.; Zheng, C.; Yin, H.; Wei, H. Network Analysis of MicroRNAs, Transcription Factors, and Target Gen Involved in Axon Regeneration. *J. Zhejiang Univ. Sci. B* **2018**, *19*, 293–304. [CrossRef]
28. Dachtler, J.; Glasper, J.; Cohen, R.N.; Ivorra, J.L.; Swiffen, D.J.; Jackson, A.J.; Harte, M.K.; Rodgers, R.J.; Clapcote, S.J. Deletion α-Neurexin II Results in Autism-Related Behaviors in Mice. *Transl. Psychiatry* **2014**, *4*, e484. [CrossRef] [PubMed]
29. Merkurjev, D.; Hong, W.-T.; Iida, K.; Oomoto, I.; Goldie, B.J.; Yamaguti, H.; Ohara, T.; Kawaguchi, S.; Hirano, T.; Martin, K.C.; et al. Synaptic N6-Methyladenosine (M6A) Epitranscriptome Reveals Functional Partitioning of Localized Transcripts. *Nat. Neuro* **2018**, *21*, 1004–1014. [CrossRef]
30. Romano, R.; Bucci, C. Role of EGFR in the Nervous System. *Cells* **2020**, *9*, 1887. [CrossRef]

1. Tavassoly, O.; Sato, T.; Tavassoly, I. Inhibition of Brain Epidermal Growth Factor Receptor Activation: A Novel Target in Neurodegenerative Diseases and Brain Injuries. *Mol. Pharmacol.* **2020**, *98*, 13–22. [CrossRef]
2. Zheng, H.; Zhang, X.; Sui, N. Advances in the Profiling of N6-Methyladenosine (M6A) Modifications. *Biotechnol. Adv.* **2020**, *45*, 107656. [CrossRef]
3. Zhao, Z.; Meng, J.; Su, R.; Zhang, J.; Chen, J.; Ma, X.; Xia, Q. Epitranscriptomics in Liver Disease: Basic Concepts and Therapeutic Potential. *J. Hepatol.* **2020**, *73*, 664–679. [CrossRef]
4. Zhu, Z.-M.; Huo, F.-C.; Pei, D.-S. Function and Evolution of RNA N6-Methyladenosine Modification. *Int. J. Biol. Sci.* **2020**, *16*, 1929–1940. [CrossRef]
5. Deng, R.; Cheng, Y.; Ye, S.; Zhang, J.; Huang, R.; Li, P.; Liu, H.; Deng, Q.; Wu, X.; Lan, P.; et al. M6A Methyltransferase METTL3 Suppresses Colorectal Cancer Proliferation and Migration through P38/ERK Pathways. *Onco. Targets Ther.* **2019**, *12*, 4391–4402. [CrossRef]
6. Luo, Q.; Rao, J.; Zhang, L.; Fu, B.; Guo, Y.; Huang, Z.; Li, J. The Study of METTL14, ALKBH5, and YTHDF2 in Peripheral Blood Mononuclear Cells from Systemic Lupus Erythematosus. *Mol. Genet. Genom. Med.* **2020**, *8*, e1298. [CrossRef]
7. Kakati, S.; Barman, B.; Ahmed, S.U.; Hussain, M. Neurological Manifestations in Systemic Lupus Erythematosus: A Single Centre Study from North East India. *J. Clin. Diagn. Res.* **2017**, *11*, OC05–OC09. [CrossRef] [PubMed]
8. Dao, F.-Y.; Lv, H.; Yang, Y.-H.; Zulfiqar, H.; Gao, H.; Lin, H. Computational Identification of N6-Methyladenosine Sites in Multiple Tissues of Mammals. *Comput. Struct. Biotechnol. J.* **2020**, *18*, 1084–1091. [CrossRef]
9. Liu, H.; Begik, O.; Lucas, M.C.; Ramirez, J.M.; Mason, C.E.; Wiener, D.; Schwartz, S.; Mattick, J.S.; Smith, M.A.; Novoa, E.M. Accurate Detection of M6A RNA Modifications in Native RNA Sequences. *Nat. Commun.* **2019**, *10*, 4079. [CrossRef] [PubMed]
10. Qiang, X.; Chen, H.; Ye, X.; Su, R.; Wei, L. M6AMRFS: Robust Prediction of N6-Methyladenosine Sites With Sequence-Based Features in Multiple Species. *Front. Genet.* **2018**, *9*, 495. [CrossRef]
11. Xiang, S.; Liu, K.; Yan, Z.; Zhang, Y.; Sun, Z. RNAMethPre: A Web Server for the Prediction and Query of MRNA M6A Sites. *PLoS ONE* **2016**, *11*, e0162707. [CrossRef]
12. Chen, W.; Feng, P.; Yang, H.; Ding, H.; Lin, H.; Chou, K.-C. IRNA-AI: Identifying the Adenosine to Inosine Editing Sites in RNA Sequences. *Oncotarget* **2016**, *8*, 4208–4217. [CrossRef] [PubMed]
13. Qiu, W.-R.; Xiao, X.; Chou, K.-C. IRSpot-TNCPseAAC: Identify Recombination Spots with Trinucleotide Composition and Pseudo Amino Acid Components. *Int. J. Mol. Sci.* **2014**, *15*, 1746–1766. [CrossRef]
14. Liu, Z.; Xiao, X.; Qiu, W.-R.; Chou, K.-C. IDNA-Methyl: Identifying DNA Methylation Sites via Pseudo Trinucleotide Composition. *Anal. Biochem.* **2015**, *474*, 69–77. [CrossRef] [PubMed]
15. Chen, W.; Tang, H.; Ye, J.; Lin, H.; Chou, K.-C. IRNA-PseU: Identifying RNA Pseudouridine Sites. *Mol. Ther. Nucleic Acids* **2016**, *5*, e332. [CrossRef]

Article

A Maximum Flow-Based Approach to Prioritize Drugs for Drug Repurposing of Chronic Diseases

Md. Mohaiminul Islam [1], Yang Wang [1] and Pingzhao Hu [1,2,3,4,*]

1 Department of Computer Science, University of Manitoba, Winnipeg, MB R3T 2N2, Canada; islammm5@myumanitoba.ca (M.M.I.); yang.wang@umanitoba.ca (Y.W.)
2 Department of Biochemistry and Medical Genetics, University of Manitoba, Winnipeg, MB R3T 2N2, Canada
3 Department of Electrical Computer Engineering, University of Manitoba, Winnipeg, MB R3T 2N2, Canada
4 CancerCare Manitoba Research Institute, Winnipeg, MB R3T 2N2, Canada
* Correspondence: pingzhao.hu@umanitoba.ca

Abstract: The discovery of new drugs is required in the time of global aging and increasing populations. Traditional drug development strategies are expensive, time-consuming, and have high risks. Thus, drug repurposing, which treats new/other diseases using existing drugs, has become a very admired tactic. It can also be referred to as the re-investigation of the existing drugs that failed to indicate the usefulness for the new diseases. Previously published literature used maximum flow approaches to identify new drug targets for drug-resistant infectious diseases but not for drug repurposing. Therefore, we are proposing a maximum flow-based protein–protein interactions (PPIs) network analysis approach to identify new drug targets (proteins) from the targets of the FDA (Food and Drug Administration) drugs and their associated drugs for chronic diseases (such as breast cancer, inflammatory bowel disease (IBD), and chronic obstructive pulmonary disease (COPD)) treatment. Experimental results showed that we have successfully turned the drug repurposing into a maximum flow problem. Our top candidates of drug repurposing, Guanidine, Dasatinib, and Phenethyl Isothiocyanate for breast cancer, IBD, and COPD were experimentally validated by other independent research as the potential candidate drugs for these diseases, respectively. This shows the usefulness of the proposed maximum flow approach for drug repurposing.

Keywords: drug–target interactions; protein–protein interactions; chronic diseases; drug repurposing; maximum flow

Citation: Islam, M.M.; Wang, Y.; Hu, A Maximum Flow-Based Approach Prioritize Drugs for Drug Repurposing of Chronic Diseases. *Life* 21, *11*, 1115. https://doi.org/ .3390/life11111115

Academic Editor: Stanislav Miertus

Received: 2 August 2021
Accepted: 18 October 2021
Published: 20 October 2021

Publisher's Note: MDPI stays neutral with regard to jurisdictional claims in published maps and institutional affiliations.

1. Introduction

Chronic diseases are usually defined as the diseases that are persistent or long-lasting and require ongoing medical attention. There are many different types of chronic diseases. For example, breast cancer starts from the breast cancer cells. However, it can also spread to other parts of the body. Breast cancer is referred to as the most frequently identified cancer in women. This is the second prominent reason for cancer death among women [1]. Of note, cancer is a multistage disease [2], increasing the mortality rate among people worldwide [3]. Several breast cancer treatment techniques are available, such as surgery, chemotherapy, radiation, and hormone therapy. Often a combination of these treatments is used in practice [4]. Other chronic diseases, such as inflammatory bowel disease (IBD) and chronic obstructive pulmonary disease (COPD), are usually consequences of many environmental and genomic factors. IBD is a chronic disease that includes both ulcerative colitis and Crohn's disease, and it lasts for a very long time. IBD results in a significant burden to our society and families. IBD triggers segments of the bowel to get red and swollen. IBD treatment involves medicines, diet modifications, and occasionally surgery [5]. The goal of such treatment options is to reduce the inflammation associated with IBD. In the long term, existing treatments may achieve reduced risks of IBD complications. COPD is a chronic lung disease that causes breathing problems. COPD is the main reason for respiratory mortality worldwide [6]. Current treatment options include lung transplants, quitting smoking, and inhalers. However, these strategies can only assist in lessening the progression of COPD. The fundamental cause of COPD is smoking [7]. Patients may not know about the disease initially, but the condition worsens over time, such as with severe breathing problems during simple tasks, e.g., walking.

There is a pressing need to identify potential drug targets and their drugs for developing personalized treatments for chronic diseases. However, new drug development takes a very long time and is extremely expensive. Usually, this type of approach takes 10–15 years and $1 billion [8]. Nevertheless, we can save time and money using old drugs for new usages called drug repurposing or repositioning. This is a helpful technique to find different indications for current medications. For example, in 2020, COVID-19 infections from the novel coronavirus became a primary worldwide public health concern [9]. As a result, it was declared as a global pandemic in 2020 [10]. The pandemic created an emergency to develop vaccines or therapeutic treatment for COVID-19 infections. However, there were no available confirmed drugs to treat COVID-19 infections. Therefore, the drug repurposing technique was used to obtain a new drug from the existing FDA-approved drugs [11,12].

There are different types of approaches to identify new indications of an FDA approved drug, such as network-based [13,14], and machine learning (ML)-based [15,16] approaches.

A biological network consists of a massive number of nodes and interactions among them. A gene can easily make a subnetwork including drug targets, and these drug targets act as the bridge between this subnetwork and the original network. We can identify the risk genes of a given disease and the associated drug targets in a biological network to remove the bridge connection between the subnetwork containing the risk genes and the original network. Therefore, we can potentially treat the disease using drugs associated with the drug targets responsible for the disease's risk genes in the network.

A network-based approach tries to find a subnetwork that provides an insight into the relationship between drugs and disease genes. For example, Cheng et al. [17] proposed a network-based system to list the drug targets using three different inference algorithms which are drug resemblance in any network, protein correspondence in any network, and recognized drug target within a bipartite network.

Yeh et al. [18] first proposed a maximum flow approach to predict a set of drugs as new effective drug targets for the treatment of prostate cancer. The idea is that the candidate proteins for a drug target with a higher flow value to the risk genes have more influence on risk genes than other candidates for the drug targets. They used microarray data [19] and an interactome (PPI) network [20] of prostate cancer to build their prediction model. Next, they used the shortest path algorithm [21] to perform a maximum flow method within their network and successfully identified 20 drug targets to reuse. These drug targets were validated using other available literature that published these same drug targets for prostate cancer.

Melak et al. [22] also used the idea of the maximum flow approach to prioritize a set of drug targets to reduce the expression of tuberculosis disease from a list of known drug targets. Yeh et al. [18] used the Pearson correlation coefficient and gene expression change between genes to calculate the weight of the edges of their PPI network. However, Melak et al. [22] used a PPI network from STRING which includes the associated weights for the edges. Thus, Yeh et al. [18] and Melak et al. [22] showed that proteins with the maximum flow to the risk genes in the PPI network could be used as targets for developing drugs treat diseases.

This study aims to apply the maximum flow technique to a PPI network with set of breast cancer, IBD, and COPD risk genes to identify new breast cancer, IBD, and COPD drugs, respectively, from a list of FDA-approved medications. We hypothesize that identifying new drugs from the existing drugs (i.e., drug repurposing) for breast cancer, IBD, and COPD can be converted into a maximum flow problem using a human interactome network (i.e., a PPI network). Furthermore, it is believed that drug targets X (proteins) connected with risk genes through a higher flow value have more impact on these risk genes than other drug targets. Therefore, these Xs can be used as potential targets for drug development for the disease's treatment. Furthermore, deletion of these Xs from the PPI network will disrupt the communication among the risk genes and proteins. Therefore, this study aims to identify a set of strongly correlated proteins with the disease risk genes from a PPI network using a maximum flow approach. Later, we can identify new candidate drugs for repurposing to treat breast cancer, IBD, and COPD associated with these targets using a drug-target interaction network.

2. Materials and Methods

2.1. Datasets

2.1.1. Protein–Protein Interaction (PPI) Network

We collected a comprehensive biological network [23] which includes 140,899 interactions among the 13,365 human proteins (genes). We used this biological network (Figure 1) to conduct our experiments.

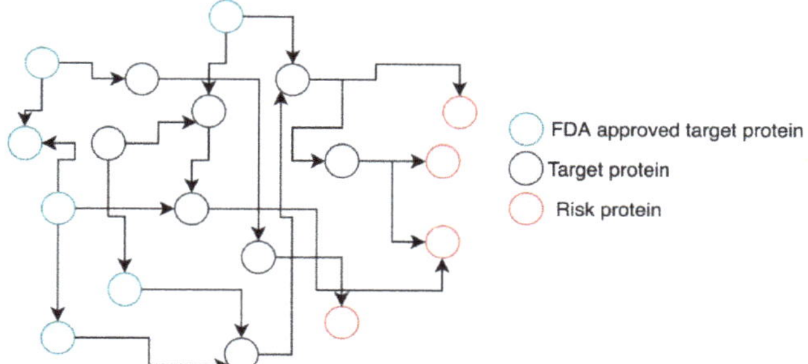

Figure 1. An example of our PPI network. The network shows the interactions among the FDA-approved drug targets (i.e., proteins), potential drug targets, and disease-associated risk proteins/genes.

2.1.2. Drug-Target Interactions (DTIs) Network

We extracted 2390 FDA-approved drug targets (DTs) in human from DrugBank [24]. However, the PPIs network described in Section 2.1.1 contains only 1926 DTs among these 2390 FDA-approved DTs. We also collected the DTIs network, which has ~13,000 DTIs among 5049 unique drugs and 3099 unique targets from the DrugBank.

2.1.3. Risk Genes

In this study, we focused our drug repurposing on the above-mentioned three diseases (breast cancer, IBD, and COPD) since they have a relatively large number of disease-specific risk genes identified from genome-wide association studies (GWAS) as described below. These risk genes make the application of the maximum flow technique to drug repurposing possible in this study. GWAS have already discovered more than 200 breast cancer risk loci. For example, Baxter et al. [25] were able to mark 63 loci and identified 110 known target genes at 33 loci. In addition, Wu et al. [26] identified 179 significant genes associated with breast cancer risk. Thus, we have collected in total 289 breast cancer risk genes from these two studies.

Previously published genomic studies identified 215 risk loci to explain the fundamental molecular biology of IBD [27]. In addition, Katrina et al. [27] marked three additional loci which have therapeutic targets in IBD. They have also prioritized 811 IBD risk genes from 240 risk variants.

A GWAS in the United Kingdom by Sakornsakolpat et al. [6] identified 82 loci associated with COPD or function. Among them, 47 loci were already known as risk loci of

COPD. Of note, Sakornsakolpat et al. [6] have identified 156 COPD risk genes from these 82 loci.

Hence, we have collected 289, 811, and 156 risk genes responsible for breast cancer, IBD, and COPD, respectively, from the earlier studies to validate the usefulness of our proposed drug repurposing method.

2.2. The Maximum Flow Algorithm for Drug Repurposing

The analysis pipeline for drug repurposing includes multiple steps, as shown in Figure 2 (taking breast cancer as an example). Below we explain the steps in more detail.

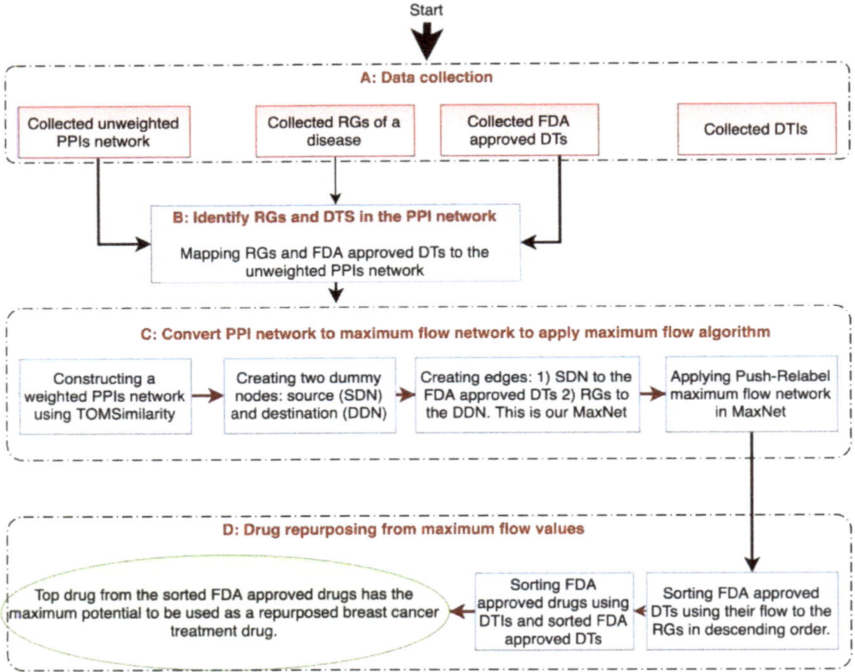

Figure 2. Analysis pipeline for the maximum flow approach to prioritize drugs for drug repurposing (taking breast cancer as an example). (**A**) Shows the types of data we collected for our experiments. (**B**) We mapped each target protein in the PPI network to be either a risk gene, FDA-approved drug target, or potential candidate target. (**C**) Shows the construction of maximum flow network from the collected PPI network to apply the Push-Relable maximum flow algorithm. (**D**) Shows the steps to repurpose existing drugs based on the maximum flow values of each target protein.

2.2.1. Constructing the Maximum Flow Network

Mapping drug targets and risk genes to the PPIs network: We first mapped the 19 FDA-approved DTs (FDA_DT) and risk genes (RGs) of a specific disease to the unweighted PPIs network (refers to a graph where edges do not have weights, and there is only one edge between any two nodes).

Constructing weighted PPIs network: We used TOMSimilarity (topological overlap matrix similarity) [19] to calculate the weight of edges between genes, and we used Equation (1) to get TOMSimilarity between two nodes in our network.

$$TOMSimilarity\,(x,y) = \frac{\left| N_{neighbor}(x) \cap N_{neighbor}(y) \right| + A_{xy}}{\min\left(\left| N_{neighbor}(x) \right|, \left| N_{neighbor}(y) \right| \right) + 1 - A_{xy}}$$

where $N_{neighbor}(x)$ is the neighbors of x,

$N_{neighbor}(y)$ is the neighbors of y,

A_{xy} is the value of the adjacency matrix (i.e., one if nodes x and y are connected and zero otherwise),

$TOMSimilarity\ (x, y)$ is the Topological Overlap Matrix Similarity between the nodes x and y.

Drug repurposing as a maximum flow problem: After the mapping of the drug targets and risk genes, we specified the drug repurposing problem into a maximum flow problem, we (1) created a dummy node SDN (i.e., the source of the network) which was connected with all the FDA_DT; (2) created another dummy node DDN (i.e., the destination of the network) which was connected with all the risk genes; (3) assigned a flow capacity (i.e., weight) using Equation (1) for each of the connections in the network. Flows in the maximum flow network follow the below rules: (1) The input flow is equal to the output flow for any node except the source and destination nodes; (2) for any edge (e) in the network, $0 \leq \text{flow}(e) \leq \text{Capacity}(e)$; (3) total flow out of the source node is equal to total flow into the destination node.

However, the connections from the dummy source to the candidate drug targets will have a dummy capacity. Each incoming edge from the dummy source node to a protein (drug target) has a capacity equal to the sum of the capacities of the outgoing edges from that protein (drug targets). Similarly, the connections from the risk genes to the dummy sink node have dummy capacities. Each outgoing edge from a risk gene to the sink node has a capacity equal to the sum of the capacities of the incoming edges to that risk gene. At this point, we had the network named MaxNet (Figure 3) to run the maximum flow algorithm.

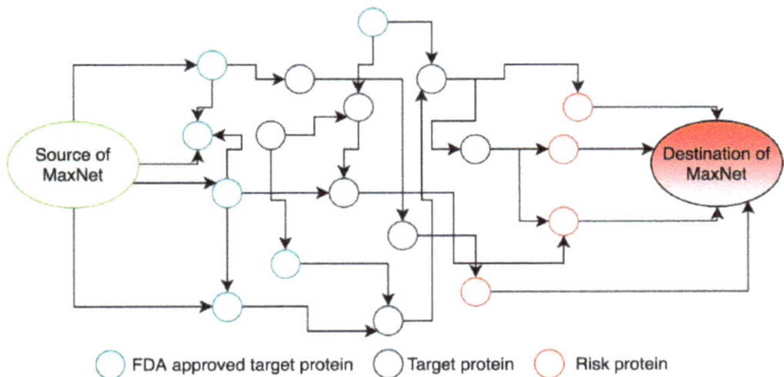

Figure 3. An example of our MaxNet.

2.2.2. Push-Relabel Maximum Flow Algorithm

We used the Push-Relabel maximum flow algorithm [28] in the MaxNet (Figure 3) to maximize the flow amount passed from the FDA-approved drug targets to the risk genes. Algorithm 1 (revised from [29]) shows the Push-Relabel maximum flow algorithm. In addition, this algorithm works with one vertex at a time. Every vertex is associated with two variables: height and excess flow. A vertex can send flows to a lower-height vertex only. The extra flow of a vertex represents the difference between the total in-flow and out-flow of that vertex. Furthermore, each edge is associated with two variables: flow (i.e., current flow through this edge) and capacity (i.e., the maximum flow we can send through this edge). This algorithm sends flows (i.e., PUSH operation) from a node (S) to its adjacent node (D) when the excess flow of D is not equal to zero and the height of D is less than the height of S. If there is no adjacent node of S with lesser height than this algorithm increases the height of S (i.e., RELABEL operation) by the minimum height of the adjacent nodes of S plus 1.

Algorithm 1 Push-Relabel_MaximumFlow_Algorithm [28].

Input: PPI, Capacity = C, N = unique nodes of PPI, start_node = SDN, destination_node = DDN.
Output: Maximum flow between SDN and DDN

    (1)   FOR i = 1 to length [N]:
        a.   HeightV [i] = 0//HeightV is height of every vertex
        b.   FlowV [i] = 0//FlowV is the flow of every vertex
    (2)   HeightV [start_node] = length [N]
    (3)   FOR i = 1 to length [PPI]:
        a.   FlowE [i] = 0//FlowE is the flow of every edge in the PPI
    (4)   V = adjacentVetex[start_node]
    (5)   FOR i = 1 to length [V]:
        a.   FlowV [V[i]] = Capacity [V[i]]
        b.   excessFlow [V[i]] = Capacity [V[i]]
    (6)   **PUSH:** FOR i = 1 to length [N]:
    If excessFlow [N[i]] ≠ 0: (in the residual graph)
           tmpV = adjacentVetex[N[i]]
           if HeightV [N[i]] > lowest_height[tmpV]
             Push_flow from N[i] to lower height vertices
    (7)   **RELABEL:** FOR i = 1 to length [N]:
    If excessFlow [N[i]] ≠ 0: (in the residual graph)
       tmpV = adjacentVetex[N[i]]
       if HeightV [N[i]] ≤ lowest_height[tmpV]
         HeightV [N[i]] = minimumHeight[tmp]

2.2.3. Drug Repurposing from Maximum Flow Values

After applying the Push-Relabel maximum flow algorithm in our MaxNet network, we sorted all the FDA drug targets into a list L_{DTs} according to their flow value to the risk genes (descending order). Then, we used this sorted list L_{DTs} of the DTs to sort the FDA-approved drugs into a list L_{drugs} using ~13,000 DTIs collected from DrugBank [29]. Hence, according to our hypothesis, the top drugs in L are the most prominent drugs that can be reused to treat the given disease associated with its risk genes.

The whole analysis pipeline of the maximum flow-based drug repurposing is summarized in Algorithm 2.

Algorithm 2 Pipeline of the maximum flow-based drug repurposing.

Input: PPI = all the PPIs, FDA_DT = all the FDA approved DTs in PPIs network, DTI = DTIs for FDA_DT, RG = risk genes, W = flow capacity of edges.
Output: CD = candidate drugs for repurposing for the treatment of breast cancer.

    1.   FOR i = 1 to length [PPI]:
        a.   Calculate flow capacity of the edge using Equation (1):
           C[i] = TOMSimilarity (PPI[i])
    2.   CREATE two dummy nodes:
        a.   source dummy node = SDN and destination dummy node = DDN
    3.   FOR i = 1 to length [FDA_DT]:
        a.   Index = length [PPI] + 1
        b.   CONNECT SDN to FDA_DT[i] and add this interaction in PPI[index]
        c.   W[index] = sum of the capacities of the outgoing edges from PPI[index]
    4.   FOR i = 1 to the length of RG:
        a.   Index = length of PPI + 1
        b.   CONNECT RG[i] to DDN and add this interaction in PPI[index]
        c.   C[index] = sum of the capacities of the incoming edges from PPI[index]
    5.   The nodes in PPIs and their associated outgoing flow value = Push-Relabel_
        MaximumFlow_Algorithm (PPI, C, SDN, DDN)
    6.   prioritized_DTs = sort the nodes in PPI in decreasing order of their outgoing flows
    7.   CD = sort drugs in DTI using prioritized_DTs

---

3. Experimental Results

3.1. Mapping Drug Targets and Disease-Specific Risk Genes to the PPIs Network

First, we collected the PPI network. This is an unweighted network. So, we calculated topological overlap similarity (TOMSimilarity) to assign weights on the edges. These weights were used as the capacities of the flow through the edges. In this PPI network, we had 1926 FDA-approved DTs. Next, we mapped disease-specific risk genes to this PPI network. The PPIs network contained 155 breast cancer RGs from the 289 breast cancer RGs identified by Baxter et al. [25] and Wu et al. [26]. It also had 565 IBD risk genes among the 811 prioritized IBD risk genes by Katrina et al. [27]. This PPI network also contained 118 COPD risk genes among the 156 COPD risk genes identified by Sakornsakolpa et al. [6]. Table 1 shows several statistical properties of the PPI network. In Table 1, transitivity refers to the probability of adjacent nodes being interconnected. It provides an intuition about the clusters in the network. Of note, in a graph, total triangles represent the total number of triangles formed by any three nodes. In addition, we also showed the PPI network's degree distribution in Figure 4. Figure 4 indicates that only a few nodes in the PPIs network have a high number of neighbors. This means the PPI network has a small number of hubs.

Table 1. Statistical properties of the PPIs network.

Properties	Values
Number of nodes	13,368
Number of edges	140,899
Transitivity	0.292
Average clustering coefficient	0.173
Edge density	0.002
Average degree	21.08
Total triangles	4,105,272

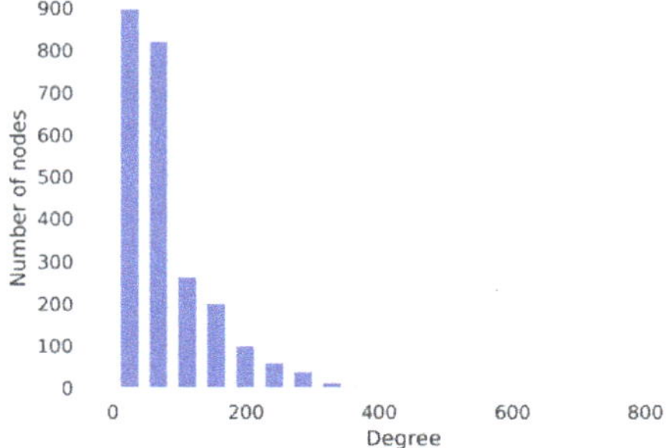

Figure 4. Degree distribution of the PPIs network.

3.2. Weights of the Interactions in PPIs Network

We calculated topological overlap similarity (TOMSimilarity) to assign weights on the edges of the unweighted PPI network. The values of these edge weights ranged from 0 to 1. We used these edge weights as flow capacity for each connection during maximum flow implementation with Algorithm 1.

3.3. Formulating Drug Repurposing as a Maximum Flow Network

FDA-approved drug targets are the network sources, while risk genes are the destinations of the network. Hence, we needed to convert this multiple sources and multiple destinations network into a single source and single destination network. To do this, we created a dummy source node and connected this node with 1926 DTs. Similarly, we created a dummy destination node and only connected this sink node with the disease-specific risk genes. As a result, there were no incoming arcs to the source node and no outgoing arcs from the destination node. We calculated the sum of the capacities of the outgoing arcs from a drug target node and put this sum as the capacities on the arcs from the dummy source node to the drug target node. Likewise, we calculated the sum of the capacities of the incoming arcs to a risk gene node and put this sum as the capacities to the arc from the risk gene node to the dummy destination node. We called this network the MaxNet.

3.4. Drug Repurposing for Breast Cancer, IBD, and COPD

We created three MaxNets (MaxNet_BC, MaxNet_IBD, and MaxNet_COPD) for breast cancer, IBD, and COPD RGs, respectively. For all three MaxNets, the dummy source node was connected with the 1926 FDA-approved DTS. However, our PPIs network contained only 155 breast cancer RGs, 565 IBD RGs, and 118 COPD RGs. Therefore, we connected the 155 breast cancer RGs with the dummy destination node in the MaxNet_BC, the 565 IBD RGs with the dummy destination node in the MaxNet_IBD, and the 118 COPD RGs with the dummy destination node in the MaxNet_COPD.

We ran the Push-Relabel maximum flow algorithm in all three MaxNets to get the maximum flow values for each node from the dummy source to the dummy destination. First, we extracted three sorted lists of the targets (FDA-approved) based on their outgoing flows from the MaxNet_BC, MaxNet_IBD, and MaxNet_COPD in descending order. Then we used these sorted lists of targets to sort the drug list using a drug-target interaction network for breast cancer, IBD, and COPD. According to our hypothesis, the top drug in each of these sorted drug lists has the maximum potential to be used as a candidate drug for the treatment of breast cancer, IBD, and COPD, respectively.

3.5. Performance Evaluation

We performed a comprehensive literature review to validate our top five repurposed candidates for breast cancer, IBD, and COPD as shown in Tables 2–4, respectively.

Table 2. The top five repurposed drugs for breast cancer.

Drug Name	Target Protein	Target Gene	Flow Value	Status	Reference
Guanidine	P78352	DLG4	0.0489	Confirmed	[30]
Phenethyl Isothiocyanate	P31946	YWHAB	0.0389	Confirmed	[31]
Caffeine	P78527	PRKDC	0.0363	Confirmed	[32]
Tamoxifen	Q05655	PRKCD	0.0363	Confirmed	[33]
(2S)-2-({6-[(3-Amino-5-chlorophenyl)amino]-9-isopropyl-9H-purin-2-yl}amino)-3-methyl-1-butanol	Q00534	CDK6	0.03319202		

Table 3. The top five repurposed drugs for IBD.

Drug Name	Target Protein	Target Gene	Flow Value	Status	Reference
Dasatinib	P12931	SRC	0.08292133	Confirmed	[34]
Phenethyl Isothiocyanate	P31946	YWHAB	0.06112281	Confirmed	[35]
Adenosine-5′	P00558	PGK1	0.04545455	Confirmed	[36]
Acetylsalicylic acid	P54646	PRKAA2	0.03627599		
Glutamic Acid	P07814	EPRS	0.03527291	Confirmed	[37]

Table 4. The top five repurposed drugs for COPD.

Drug Name	Target Protein	Target Gene	Flow Value	Status	Reference
Phenethyl Isothiocyanate	P31946	YWHAB	0.05054656	Confirmed	[38]
Minocycline	P42574	CASP3	0.03767546	Confirmed	[39]
Pseudoephedrine	P15336	ATF2	0.03201844	Confirmed	[38]
Methyl 4,6-O-[(1R)-1-carboxyethylidene]-beta-D-galactopyranoside	P02743	APCS	0.03150388		
NADH	O43920	NDUFS5	0.02409639	Confirmed	[40]

In addition, we have shown the top 10 prioritized repurposed drugs in the Supplementary Tables S1–S3 for each of these diseases.

3.6. Performance Comparison with Other Methods

We used the same datasets to compare the performance of our maximum flow-based drug prioritization with the baseline methods, such as degree, betweenness centrality, closeness centrality, random walk, and page rank (Table 5). Degree centrality refers to the number of incoming links to a node and ranks the risk genes by their degree value. Closeness centrality is defined as the geodesic distance (normalized) for any node to any other node in the network. Finally, the betweenness centrality of a node denotes the number of shortest paths that include this node. First, we used MATLAB functions to calculate degree centrality, closeness centrality, and betweenness centrality from the PPI network for each disease of interest (breast cancer, IBD, and COPD). Then, we sorted each of these lists of targets in descending order. Furthermore, we obtained a sorted list of candidate drugs using these sorted targets and a drug-target interaction dataset. Then we used the python functions of random walk [41] and page rank [42] to calculate the importance of each target associated with breast cancer, IBD, and COPD in the PPI network. Finally, we used sorted random walk [41] and page rank [42] (descending order) lists of targets to identify potential drug repurposing candidates from the drug-target interaction network we collected from the DrugBank database.

Table 5. Number of confirmed disease-specific candidates by the baseline approaches for drug repurposing in the list of top five candidate drugs.

Method	Number of Confirmed Candidates in Top 5 Candidate Drug List		
	Breast Cancer	IBD	COPD
Degree centrality	0	0	1
Closeness centrality	2	1	0
Betweenness centrality	0	0	2
Random walk [41]	0	2	2
Page rank [42]	2	2	2
Our proposed framework	4	4	4

4. Discussion

Traditional machine learning methods, such as naive Bayesian, support vector machines, and the latest deep neural networks, reveal their effectiveness for drug discovery. Zhao et al. proposed a method that uses drug-induced expression profiles to predict the sign of a disease in psychiatry [43]. Saberian et al. [44] introduced a framework that takes anti-similarity between drugs and a disease as input to train a model. Their model can predict new usage apart from the primary indications of a drug. However, researchers have concerns about using conventional machine learning techniques for this purpose because of the background noisiness and the high-dimensionality nature of the biological data [45]. Hence, Cheng et al. [17] used a chemical structure with the genome sequence to perform the drug and protein resemblance checking. At the same time, they anticipated related drugs might share identical drug targets for a disease. However, they did not find any helpful result from these similarities checking among the drugs. Nevertheless, they concluded that the chemical structure could not be represented as a parameter to identify similar drugs or proteins. Estrada et al. [46] also used a biological network's global measure such as closeness/betweenness centrality to identify drug targets. They considered a node in the network as the drug target if it has a higher closeness/betweenness centrality value than the other nodes. These measures are based on the shortest paths in the network. In addition, random walk [41] and page rank (the algorithm that Google uses for their search engine) [42] can be used to extract such global measures to identify potential new drug targets. In this study, we adopted a maximum flow-based approach similar to Yeh et al. [18] and Melak et al. [22] to prioritize FDA-approved drugs repurposed for breast cancer, IBD and COPD.

We used a PPI network [23] to conduct our experiments. The investigators mentioned that these interactions do not contain any interactions estimated from gene expression data. These interactions fall into the following categories: protein–protein interactions (most of the interactions fall into this category), regulatory interactions, protein database, and signaling interactions [47]. However, this PPI network is not weighted. Therefore, we converted our PPI network to a weighted network using TOMSimilarity. We used TOM Similarity because Langfelder et al. [48] showed its effectiveness as a highly robust measure of network interconnectedness (proximity) for the hierarchical clustering of biological data. TOMSimilarity calculates the topological similarity between two connected proteins (i.e genes) using an adjacency matrix. Then, we applied the Push-Relabel algorithm to obtain the node importance based on its outflow. This algorithm works locally rather than looking into the entire residual graph (this graph indicates if it is possible to send flows from the source to the destination of the network) to find an augmenting path to send flows.

The primary usage of our most promising candidate drug, "Guanidine" (Table 2), is to treat muscle weakness caused by Eaton-Lambert syndrome. In 2009, Meruling et al. [30] showed that at 0.5 microM, dextran aminoguanidine conjugate killed more than 95% of the breast cancer cells compared to 25% for Adriamycin. The second candidate, "Phenethyl Isothiocyanate" (PEITC) (Table 2), with unique specificity, has promising results for HER breast cancer patients. "Caffeine" (Table 2) is primarily used to restore mental alertness when fatigue or drowsiness are present and for the treatment of post-dural lumbar puncture headaches. However, Pantziarka et al. [32] confirmed that caffeine could be used to treat breast cancer. The fourth candidate, "Tamoxifen," is primarily used for breast cancer. Hence, we showed the top five candidate drugs using our proposed framework in Table.

According to our proposed framework, the most promising candidate drug used as the IBD repurposed drug is "Dasatinib" (Table 3). It has been shown that Dasatinib is helpful to decrease the inflammation in a rodent model of colitis [34] for ulcerative colitis type IBD. Therefore, the study concluded that Dasatinib could be a potential candidate for ulcerative colitis treatment. Our second IBD repurposed drug candidate is "Phenethyl Isothiocyanate" (PEITC) (Table 3). PEITC Essential Oil contains more than 95% of PEITC. Therefore, De et al. [35] confirmed PEITC essential oil as a potential treatment for ulcerative colitis patients. The third candidate, "Adenosine" (Table 3), is working as a modulator for inflammation (including Crohn's disease and ulcerative colitis) both in humans and animals [36]. Our last candidate, "Glutamic Acid" (Table 3), was confirmed by [37] as an amino acid is an adjuvant ulcerative colitis type of IBD treatment. Furthermore, the investigators showed that microinjection of this amino acid into the paraventricular nucleus on ulcerative colitis in rats significantly improved anti-oxidation levels. This outcome suggests that glutamic acid is a potential candidate for a therapeutic application of paraventricular nucleus regulation in ulcerative colitis. However, the doses of glutamic acid may change for the naturally-occurring IBD.

The primary usage of Phenethyl Isothiocyanate (PEITC) is the treatment of lung cancer [38]. Nonetheless, our proposed framework considers this drug the most favorable contender in our top five candidate drugs list (Table 4) for COPD. Our next candidate

"Minocycline" (Table 4), is effective as an addition to treatment with cyclophosphamide in reducing the number of lung cancer [39]. The third candidate, "Pseudoephedrine" (Table 4) can also be used for COPD-related diseases such as the treatment of nasal and sinus congestion that is caused by a breathing illness (e.g., bronchitis) [49,50]. Finally, the last candidate, "NADH" (Table 4), improves trial COPD [40], emphasizing a probable helpful treatment for COPD.

From Table 5, it is self-evident that our proposed framework outperformed baseline methods (degree centrality, betweenness and closeness centrality, random walk [15], and page rank [42]) in prioritizing drug candidates for disease-specific drug repurposing. A literature review-based validation confirmed that our proposed framework correctly prioritized four out of the top five candidate drugs for drug repurposing for breast cancer, IBD, and COPD, respectively. On the other hand, degree and betweenness centrality methods have only one and two confirmed drug candidates, respectively, to be used as repurposed drugs for COPD only. Closeness centrality has two and one confirmed drug candidates as repurposed drugs for breast cancer and IBD, respectively. The random walk has zero, two, and two confirmed drugs in the predicted top five drugs to treat breast cancer, IBD, and COPD diseases, respectively. However, the page rank approach listed two confirmed drugs for each disease in the top five predicted lists of drugs.

The above literature review-based comparison suggests that our proposed framework can be used for novel drug discovery and drug repurposing. Therefore, it may be promising to use the proposed drug repurposing framework to prioritize candidate disease-specific repurposed drugs and disease-specific primary drugs.

5. Conclusions

This study aims to formulate drug repurposing for a specific disease as a maximum flow problem. We used a human interactome network and a set of FDA-approved drug targets along with different disease-specific (breast cancer, IBD, and COPD) risk genes to perform our experiments. We hypothesized that our proposed framework would identify a set of FDA-approved drugs that can be repurposed to treat breast cancer, IBD, and COPD. Experimental results showed that we had identified a prioritized list of drug targets and associated drugs that can be reused to treat these diseases. Furthermore, our proposed framework identified the natural flow to strongly influence the disease genes without any prior knowledge. Finally, we performed a comprehensive literature review to validate our proposed framework's performance. This validation shows that our proposed framework outperformed other baseline methods regarding the total number of confirmed repurposed drugs. The validation also suggests that our drug repurposing approach can also be used for novel drug discovery.

Future works of this study include experiments and clinical trials with our prioritized lists of candidate drugs. These approaches will confirm whether our candidate drugs have the potential to treat breast cancer, IBD, and COPD, respectively.

Supplementary Materials: The following are available online at https://www.mdpi.com/article/10.3390/life11111115/s1, Table S1: The top 10 prioritized repurposed drugs for breast cancer, Table S2: The top 10 prioritized repurposed drugs for IBD, Table S3: The top 10 prioritized repurposed drugs for COPD.

Author Contributions: Conceptualization, M.M.I. and P.H.; methodology, M.M.I.; software, M.M.I.; validation, M.M.I.; formal analysis, M.M.I.; investigation, M.M.I.; resources, P.H.; data curation, P.H.; writing—original draft preparation, M.M.I.; writing—review and editing, P.H.; visualization, M.M.I.; supervision, P.H. and Y.W.; project administration, P.H.; funding acquisition, P.H. and Y.W. All authors have read and agreed to the published version of the manuscript.

Funding: This work is partially funded by a grant from the University of Manitoba Collaborative Research Program and a grant from the Natural Sciences and Engineering Research Council of Canada (NSERC) (EGP-543968-2019).

Institutional Review Board Statement: Ethical review and approval were waived for this study because all the data used in the study are publicly available.

Informed Consent Statement: Not applicable.

Data Availability Statement: All the datasets used for the experimental analysis are publicly available

Acknowledgments: We thank Mark Alexiuk very much for his excellent supervisionfor Md. Mohaiminul Islam, and the kind support from Sightline Innovation on the project.

Conflicts of Interest: The authors declare no conflict of interest.

References

1. National Breast Cancer Foundation. Available online: https://www.nationalbreastcancer.org/ (accessed on 3 September 2021
2. Vogelstein, B.; Kinzler, K.W. Cancer Genes and the Pathways They Control. *Nat. Med.* **2004**, *10*, 789–799. [CrossRef]
3. Cancer Statistics—National Cancer Institute. Available online: https://www.cancer.gov/about-cancer/understanding/statistic (accessed on 3 September 2021).
4. Breast Cancer: Types of Treatment | Cancer.Net. Available online: https://www.cancer.net/cancer-types/breast-cancer/type treatment (accessed on 10 October 2021).
5. Inflammatory Bowel Disease. Available online: https://kidshealth.org/en/teens/ibd.html (accessed on 3 September 2021).
6. Sakornsakolpat, P.; Prokopenko, D.; Lamontagne, M.; Reeve, N.F.; Guyatt, A.L.; Jackson, V.E.; Shrine, N.; Qiao, D. Expande Genetic Landscape of Chronic Obstructive Pulmonary Disease Reveals Heterogeneous Cell Type and Phenotype Association *BioRxiv* **2018**, 355644. [CrossRef]
7. Chornic Obstructive Pulmonary Disease. Available online: https://lung.ca/copd (accessed on 28 August 2021).
8. Lipman, A.G. Drug Repurposing and Repositioning: Workshop Summary. *J. Pain Palliat. Care Pharmacother.* **2015**, *2* 81. [CrossRef]
9. Hui, D.S.; Azhar, E.I.; Madani, T.A.; Ntoumi, F.; Kock, R.; Dar, O.; Ippolito, G.; Mchugh, T.D.; Memish, Z.A.; Drosten, C.; et The Continuing 2019-NCoV Epidemic Threat of Novel Coronaviruses to Global Health—The Latest 2019 Novel Coronavir Outbreak in Wuhan, China. *Int. J. Infect. Dis.* **2020**, *91*, 264–266. [CrossRef]
10. Wang, C.; Horby, P.W.; Hayden, F.G.; Gao, G.F. A Novel Coronavirus Outbreak of Global Health Concern. *Lancet* **2020**, *3*: 470–473. [CrossRef]
11. Singhal, T. A Review of Coronavirus Disease-2019 (COVID-19). *Indian J. Pediatr.* **2020**, *87*, 281. [CrossRef] [PubMed]
12. Rabi, F.A.; Al Zoubi, M.S.; Kasasbeh, G.A.; Salameh, D.M.; Al-Nasser, A.D. SARS-CoV-2 and Coronavirus Disease 2019: What Know So Far. *Pathogens* **2020**, *9*, 231. [CrossRef]
13. Yu, L.; Huang, J.; Ma, Z.; Zhang, J.; Zou, Y.; Gao, L. Inferring Drug-Disease Associations Based on Known Protein Complex *BMC Med. Genom.* **2015**, *8*, 1–3. [CrossRef] [PubMed]
14. Zhou, Y.; Hou, Y.; Shen, J.; Huang, Y.; Martin, W.; Cheng, F. Network-Based Drug Repurposing for Novel Coronavirus 20 NCoV/SARS-CoV-2. *Cell Discov.* **2020**, *6*, 1–8. [CrossRef] [PubMed]
15. Zhang, M.; Schmitt-Ulms, G.; Sato, C.; Xi, Z.; Zhang, Y.; Zhou, Y.; George-Hyslop, P.S.; Rogaeva, E. Drug Repositioning Alzheimer's Disease Based on Systematic "omics" Data Mining. *PLoS ONE* **2016**, *11*, e0168812. [CrossRef] [PubMed]
16. Rodriguez, S.; Hug, C.; Todorov, P.; Moret, N.; Boswell, S.A.; Evans, K.; Zhou, G.; Johnson, N.T.; Hyman, B.T.; Sorger, P.K.; e Machine Learning Identifies Candidates for Drug Repurposing in Alzheimer's Disease. *Nat. Commun.* **2021**, *12*, 1– [CrossRef] [PubMed]
17. Cheng, F.; Liu, C.; Jiang, J.; Lu, W.; Li, W.; Liu, G.; Zhou, W.; Huang, J.; Tang, Y. Prediction of Drug-Target Interactions and Dr Repositioning via Network-Based Inference. *PLoS Comput. Biol.* **2012**, *8*, e1002503. [CrossRef] [PubMed]
18. Yeh, S.H.; Yeh, H.Y.; Soo, V.W. A Network Flow Approach to Predict Drug Targets from Microarray Data, Disease Genes a Interactome Network—Case Study on Prostate Cancer. *J. Clin. Bioinforma.* **2012**, *2*, 1–11. [CrossRef] [PubMed]
19. Demeter, J.; Beauheim, C.; Gollub, J.; Hernandez-Boussard, T.; Jin, H.; Maier, D.; Matese, J.C.; Nitzberg, M.; Wymore, Zachariah, Z.K.; et al. The Stanford Microarray Database. *Nucleic Acids Res.* **2001**, *29*, 152–155. [CrossRef]
20. Lapointe, J.; Li, C.; Higgins, J.P.; van de Rijn, M.; Bair, E.; Montgomery, K.; Ferrari, M.; Egevad, L.; Rayford, Bergerheim, U.; et al. Gene Expression Profiling Identifies Clinically Relevant Subtypes of Prostate Cancer. *Proc. N Acad. Sci. USA* **2004**, *101*, 811–816. [CrossRef] [PubMed]
21. Floyd, R.W. Algorithm 97: Shortest Path. *Commun. ACM* **1962**, *5*, 345. [CrossRef]
22. Melak, T.; Gakkhar, S. Maximum Flow Approach to Prioritize Potential Drug Targets of Mycobacterium Tuberculosis H3: from Protein-Protein Interaction Network. *Clin. Transl. Med.* **2015**, *4*, 1–10. [CrossRef]
23. Menche, J.; Sharma, A.; Kitsak, M.; Ghiassian, S.D.; Vidal, M.; Loscalzo, J.; Barabási, A.L. Uncovering Disease-Disease Relat ships through the Incomplete Interactome. *Science* **2015**, *347*, 1257601. [CrossRef]
24. Wishart, D.S.; Feunang, Y.D.; Guo, A.C.; Lo, E.J.; Marcu, A.; Grant, J.R.; Sajed, T.; Johnson, D.; Li, C.; Sayeeda, Z.; et al. DrugB 5.0: A Major Update to the DrugBank Database for 2018. *Nucleic Acids Res.* **2018**, *46*, D1074–D1082. [CrossRef]

25. Baxter, J.S.; Leavy, O.C.; Dryden, N.H.; Maguire, S.; Johnson, N.; Fedele, V.; Simigdala, N.; Martin, L.A.; Andrews, S.; Wingett, S.W.; et al. Capture Hi-C Identifies Putative Target Genes at 33 Breast Cancer Risk Loci. *Nat. Commun.* **2018**, *9*, 1–3. [CrossRef]

26. Wu, L.; Shi, W.; Long, J.; Guo, X.; Michailidou, K.; Beesley, J.; Bolla, M.K.; Shu, X.O.; Lu, Y.; Cai, Q.; et al. A Transcriptome-Wide Association Study of 229,000 Women Identifies New Candidate Susceptibility Genes for Breast Cancer. *Nat. Genet.* **2018**, *50*, 968–978. [CrossRef]

27. De Lange, K.M.; Moutsianas, L.; Lee, J.C.; Lamb, C.A.; Luo, Y.; Kennedy, N.A.; Jostins, L.; Rice, D.L.; Gutierrez-Achury, J.; Ji, S.G.; et al. Genome-Wide Association Study Implicates Immune Activation of Multiple Integrin Genes in Inflammatory Bowel Disease. *Nat. Genet.* **2017**, *49*, 256–261. [CrossRef]

28. Goldberg, A.V.; Tarjan, R.E. A New Approach to the Maximum-Flow Problem. *J. ACM* **1988**, *35*, 921–940. [CrossRef]

29. Wishart, D.S.; Knox, C.; Guo, A.C.; Cheng, D.; Shrivastava, S.; Tzur, D.; Gautam, B.; Hassanali, M. DrugBank: A Knowledgebase for Drugs, Drug Actions and Drug Targets. *Nucleic Acids Res.* **2008**, *36* (Suppl. S1), 901–906. [CrossRef]

30. Meurling, L.; Marquez, M.; Nilsson, S.; Holmberg, A.R. Polymer-Conjugated Guanidine Is a Potentially Useful Anti-Tumor Agent. *Int. J. Oncol.* **2009**, *35*, 281–285. [CrossRef] [PubMed]

31. Gupta, P.; Srivastava, S.K. Antitumor Activity of Phenethyl Isothiocyanate in HER2-Positive Breast Cancer Models. *BMC Med.* **2012**, *10*, 1–8. [CrossRef] [PubMed]

32. Pantziarka, P.; Sukhtame, V.; Meheus, L.; Sukhatme, V.P.V.V.; Bouche, G.; Meheus, L.; Sukhatme, V.P.V.V.; Bouche, G. Repurposing Non-Cancer Drugs in Oncology—How Many Drugs Are out There? *bioRxiv* **2017**, 197434. [CrossRef]

33. Tamoxifen. Available online: https://www.webmd.com/drugs/2/drug-4497/tamoxifen-oral/details (accessed on 29 August 2021).

34. Can, G.; Ayvaz, S.; Can, H.; Karaboğa, İ.; Demirtaş, S.; Akşit, H.; Yılmaz, B.; Korkmaz, U.; Kurt, M.; Karaca, T. The Efficacy of Tyrosine Kinase Inhibitor Dasatinib on Colonic Mucosal Damage in Murine Model of Colitis. *Clin. Res. Hepatol. Gastroenterol.* **2016**, *40*, 504–516. [CrossRef] [PubMed]

35. Dey, M.; Kuhn, P.; Ribnicky, D.; Premkumar, V.; Reuhl, K.; Raskin, I. Dietary Phenethylisothiocyanate Attenuates Bowel Inflammation in Mice. *BMC Chem. Biol.* **2010**, *10*, 1–2. [CrossRef]

36. Ye, J.H.; Rajendran, V.M. Adenosine: An Immune Modulator of Inflammatory Bowel Diseases. *World J. Gastroenterol.* **2009**, *15*, 4491. [CrossRef] [PubMed]

37. Li, T.T.; Zhang, J.F.; Fei, S.J.; Zhu, S.P.; Zhu, J.Z.; Qiao, X.; Liu, Z.B. Glutamate Microinjection into the Hypothalamic Paraventricular Nucleus Attenuates Ulcerative Colitis in Rats. *Acta Pharmacol. Sin.* **2014**, *35*, 185–194. [CrossRef]

38. Drugbank. Available online: https://www.drugbank.ca/drugs/DB12695 (accessed on 2 September 2021).

39. Sotomayor, E.A.; Teicher, B.A.; Schwartz, G.N.; Holden, S.A.; Menon, K.; Herman, T.S.; Frei, E. Minocycline in Combination with Chemotherapy or Radiation Therapy in Vitro and in Vivo. *Cancer Chemother. Pharmacol.* **1992**, *30*, 377–384. [CrossRef]

40. Li, X.; Yang, H.; Sun, H.; Lu, R.; Zhang, C.; Gao, N.; Meng, Q.; Wu, S.; Wang, S.; Aschner, M.; et al. Taurine Ameliorates Particulate Matter-Induced Emphysema by Switching on Mitochondrial NADH Dehydrogenase Genes. *Proc. Natl. Acad. Sci. USA* **2017**, *114*, E9655–E9664. [CrossRef] [PubMed]

41. Newman, M.E.J. A Measure of Betweenness Centrality Based on Random Walks. *Soc. Netw.* **2005**, *27*, 39–54. [CrossRef]

42. Li, Y. Toward a Qualitative Search Engine. *IEEE Internet Comput.* **1998**, *2*, 24–29. [CrossRef]

43. Zhao, K.; So, H.-C. A Machine Learning Approach to Drug Repositioning Based on Drug Expression Profiles: Applications to Schizophrenia and Depression/Anxiety Disorders. *arXiv* **2017**, arXiv:1706.03014.

44. Saberian, N.; Peyvandipour, A.; Donato, M.; Ansari, S.; Draghici, S. A New Computational Drug Repurposing Method Using Established Disease–Drug Pair Knowledge. *Bioinformatics* **2019**, *35*, 3672–3678. [CrossRef]

45. Napolitano, F.; Zhao, Y.; Moreira, V.M.; Tagliaferri, R.; Kere, J.; D'Amato, M.; Greco, D. Drug Repositioning: A Machine-Learning Approach through Data Integration. *J. Cheminform.* **2013**, *5*, 1–9. [CrossRef]

46. Estrada, E. Protein Bipartivity and Essentiality in the Yeast Protein-Protein Interaction Network. *J. Proteome Res.* **2006**, *5*, 2177–2184. [CrossRef]

47. Ruepp, A.; Waegele, B.; Lechner, M.; Brauner, B.; Dunger-Kaltenbach, I.; Fobo, G.; Frishman, G.; Montrone, C.; Mewes, H.W. CORUM: The Comprehensive Resource of Mammalian Protein Complexes-2009. *Nucleic Acids Res.* **2009**, *38* (Suppl. S1), 497–501. [CrossRef]

48. Langfelder, P.; Zhang, B.; Horvath, S. Defining Clusters from a Hierarchical Cluster Tree: The Dynamic Tree Cut Package for R. *Bioinformatics* **2008**, *124*, 719–720. [CrossRef]

49. Pseudoephedrine. Available online: https://go.drugbank.com/drugs/DB00852 (accessed on 28 August 2021).

50. Sudafed Oral: Uses, Side Effects, Interactions, Pictures, Warnings & Dosing—WebMD. Available online: https://www.webmd.com/drugs/2/drug-6573/sudafed-oral/details (accessed on 28 August 2021).

Article

Electrocardiogram Quality Assessment with a Generalized Deep Learning Model Assisted by Conditional Generative Adversarial Networks

ue Zhou [1], Xin Zhu [1,*], Keijiro Nakamura [2,*] and Mahito Noro [3]

[1] Biomedical Information Engineering Lab, The University of Aizu, Aizu-Wakamatsu, Fukushima 965-8580, Japan; d8212108@u-aizu.ac.jp
[2] Division of Cardiovascular Medicine, Toho University Ohashi Medical Center, Tokyo 153-8515, Japan
[3] Division of Cardiovascular Medicine, Odawara Cardiovascular Hospital, Tokyo 250-0873, Japan; noro@ojh.or.jp
[*] Correspondence: zhuxin@u-aizu.ac.jp (X.Z.); keijiro.nakanmura@med.toho-u.ac.jp (K.N.); Tel.: +81-242-37-2771 (X.Z.); +81-3-468-1251 (K.N.)

Abstract: The electrocardiogram (ECG) is widely used for cardiovascular disease diagnosis and daily health monitoring. Before ECG analysis, ECG quality screening is an essential but time-consuming and experience-dependent work for technicians. An automatic ECG quality assessment method can reduce unnecessary time loss to help cardiologists perform diagnosis. This study aims to develop an automatic quality assessment system to search qualified ECGs for interpretation. The proposed system consists of data augmentation and quality assessment parts. For data augmentation, we train a conditional generative adversarial networks model to get an ECG segment generator, and thus to increase the number of training data. Then, we pre-train a deep quality assessment model based on a training dataset composed of real and generated ECG. Finally, we fine-tune the proposed model using real ECG and validate it on two different datasets composed of real ECG. The proposed system has a generalized performance on the two validation datasets. The model's accuracy is 97.1% and 96.4%, respectively for the two datasets. The proposed method outperforms a shallow neural network model, and also a deep neural network models without being pre-trained by generated ECG. The proposed system demonstrates improved performance in the ECG quality assessment, and it has the potential to be an initial ECG quality screening tool in clinical practice.

Keywords: data augmentation; deep learning; ECG quality assessment

ation: Zhou, X.; Zhu, X.; kamura, K.; Noro, M. ctrocardiogram Quality sessment with a Generalized Deep rning Model Assisted by nditional Generative Adversarial works. *Life* **2021**, *11*, 1013. s://doi.org/10.3390/life11101013

demic Editors: Md. af-Ul-Amin, Shigehiko Kanaya, aki Ono and Ming Huang

eived: 31 August 2021
epted: 21 September 2021
lished: 26 September 2021

lisher's Note: MDPI stays neutral regard to jurisdictional claims in ished maps and institutional affil- ns.

1. Introduction

Electrocardiogram (ECG) is widely used for cardiovascular disease diagnosis, treatment, and daily personal health monitoring via wearable devices [1,2]. ECG signals are expected to have sufficient signal quality to extract temporal and morphological information for further analysis, such as heart rate variability (HRV) analysis and arrhythmia classification [3,4]. Low-quality ECG signals owing to baseline wander, muscle artifacts, and power-line interferences may cause false ECG arrhythmia alarms [5]. Additionally, ECG collected by wearable devices may include severe electrode motion artifacts, plain lines, and huge impulses due to lead-off. In particular, electrode motion artifacts may be treated as ectopic beats and cannot be removed by simple filters. This is one of the major factors that cause alarm fatigue [6–8]. In clinical practice, before disease diagnosis, low-quality ECG signals are expected to be removed through manual screening by technicians. However, manual quality screening is time-consuming, laborious, and experience-dependent. Therefore, a reliable automatic ECG signal quality assessment system is significant for ECG technicians and cardiologists.

To date, many studies have been conducted on ECG quality assessment. PhysioNet organized a challenge in cardiology in 2011 to classify 12-lead ECG signals as acceptable or unacceptable [6,9]. Quesnel et al. evaluated the quality of ECG signals contaminated with various levels of motion artifacts. They segmented PQRST complexes, which were aligned and averaged to form an estimate of true PQRST complexes. Then, a signal-to noise ratio (SNR) was estimated by comparing each PQRST complex to the average PQRST complex. In this way, they got a 0.89 Pearson correlation coefficient between estimated and real SNRs [10]. The machine learning technique was also implemented in the ECG quality assessment. Redmond et al. used a Parzen window classifier to classify noisy and clean ECG, and got 82% and 78.7% accuracies using human and automatic annotation features, respectively [11]. Shahriari et al. obtained ECG signals from an ECG alarm study at the University of California, San Francisco (UCSF) and PhysioNet Computing in Cardiology Challenge 2011. They developed an image-based ECG quality assessment method. They computed a structural similarity measure (SSIM) at first, and then selected representative ECG images from the training dataset as templates. The SSIM between each ECG image and all the templates were used to build the features and input them into linear discriminant analysis classifier. The classifier achieved 93.1% and 82.5% accuracies in the UCSF and Cardiology Challenge 2011 database, respectively [12]. Zhao et al. manually extracted six features, such as R peaks, the power spectrum distribution of QRS complexes and so forth to build fuzzy vectors. They used the fuzzy comprehensive evaluation method as a feature analysis module. Their model demonstrated a 94.67% accuracy, 90.33% recall, and 93.00% specificity, training and testing on data from the PhysioNet computing in Cardiology Challenge 2011 and 2017 [13]. In 2019, Moeyersons et al. used data from a sleep study collected by the University Hospital Leuven, PhysioNet Computing in Cardiology Challenge 2017 and MIT-BIH Noise Stress Test Database with manual labels. They segmented the ECG signal into 5 s episodes after filtering. Each episode was characterized by an autocorrelation function, and then three features were extracted and fed to a RUSBoot classifier. For Challenge 2017 and Sleep Study Datasets, they obtained a recall of 79.4% and 96.6%, specificity of 78.7% and 84.8%, and area under the curve of 0.928 and 0.970, respectively [14]. More recently, Fu et al. assessed the quality of wearable ECG signals collected via Lenovo H3 Devices. They compared three machine learning algorithms: the support vector machine (SVM), least-squares SVM (LS-SVM), and long short-term memory (LSTM) with manually extracted features. The LSTM models achieved the best performance with 95.5% accuracy [15].

The above studies usually follow three procedures. The first procedure is signal prepossessing, such as filtering. Then, feature extraction is the most important step, and directly affects the model's performance. However, no gold standard exists to identify necessary, effective, or redundant features. As a result, feature extraction usually depends on the experiences of researchers. The final step is model development using machine learning techniques or designing decision rules via setting thresholds or computing related statistical values based on extracted features. However, feature extraction, decision rules or threshold-making are experience-dependent, and it is hard to cover or find out all significant features, such as QRS-related information [11,13], R-peak-related information [12] or autocorrelation function-related features [14]. In addition, significant features may vary with decision strategies and models. Furthermore, features manually extracted from certain dataset may not be generalized on other datasets. For example, Shahriari et al. manually extracted the same features on two datasets: the UCSF and Cardiology Challenge 2011 database, but their model had a considerable difference of performance on another two datasets (UCSF vs. Cardiology Challenge 2011: accuracy: 93.1% vs. 82.5%, sensitivity: 96.3% vs. 83.9%, specificity: 90.0% vs. 77.7%) [12]. Manually extracted features generalized for different datasets are usually impractical in view of the costly and limited medical databases.

In this study, the proposed ECG quality assessment system consists of two stages: data augmentation using adversarial networks, and quality assessment using deep neural networks. The goal of data augmentation is to generate versatile ECG to improve training efficiency. The proposed system can automatically extract features from a

ECG signals and make final decisions. In this case, the system can avoid relying on experience- and database-specific manual features for model development, and thresholds or rules for decision-making; therefore, they may have better generalization ability. The system demonstrates improved performance on two different datasets, and outperforms the shallow neuronal networks model and deep neural networks model without data augmentation. All the experiments were conducted using MATLAB R2019b [16] and TensorFlow 2.3.0. [17].

2. Materials and Methods

2.1. Datasets Introduction and Construction

This study uses data from PhysioNet Computing in the Cardiology Challenge 2017 (PCCC2017) database [18], TELE ECG database [19], MIT-BIH arrhythmia database (MIT-BIHA) [20], and MIT-BIH normal sinus rhythm database (MIT-BIHNSR) [21].

PCCC2017 aims to classify single-lead ECG recordings to the sinus rhythm, atrial fibrillation (AF), alternative rhythm, or as too noisy. All the recordings last for 9 to 60 s, sampled at 300 Hz. Then, each recording was resampled to 500 Hz and segmented into segments of 10 s duration with 2 s and 4 s overlap, respectively, to increase the number of unacceptable segments (in the noisy category). In total, there are 555 unacceptable and select 2618 acceptable ECG segments from this dataset, each with 10 s duration.

The TELE ECG database was initially recorded by Redmond et al. [11] from 288 home-dwelling patients. Each ECG recording was collected with single-lead and sampled at 500 Hz. Khamis et al. regarded this database as poor-quality telehealth ECG [22]. In this study, all the recordings are marked as noisy ECG signals as well. After signal segmentation, there are 734 unacceptable 10 s ECG segments.

The PCCC2017 and TELE ECG database officially provided specific quality labels for ECG segments, and then the two databases are combined to a new dataset named as COMD. COMD consists of 1289 unacceptable and 2618 acceptable ECG segments in total.

MIT-BIHA contains 48 two-channel ECG recordings, each with a duration of 30 min, sampled at 360 Hz. MIT-BIHNSR includes 18 long-term ECG recordings, and all the recordings are sampled at 128 Hz. The MIT-BIH noise stress test database (MIT-BIHNST) was created by using two clean ECG recordings (118 and 119) from the MIT-BIHA and adding noise recordings on them [7]. The noise recordings are available in MIT-BIHNST, including baseline wander (bm), muscle artifact (ma), and electrode motion artifact (em). Inspired by the construction method of MIT-BIHNST, the original ECG recordings in MIT-BIHA and MIT-BIHNSR as regarded clean signals (acceptable recordings) as well in this study, and then the same noise-added rules as MIT-BIHNST are followed to recreate a new noise-included dataset (RECD for short) using a WFDB software package [23]. The recreated ECG recordings will include severe "ma", "bm", or "em" noises provided by MIT-BIHNST or a Gaussian noise and power interference simulated by MATLAB. The detailed noise-added rule is listed in Table 1, where "g" means Gaussian noise generated by a MATLAB function "awgn" in the WFDB software package, and "p" means a power interference simulated by a sine function, with 60 Hz frequency. To increase data diversity, RECD is created by complying with different noise combinations. Then, all the ECG recordings in RECD are resampled at 500 Hz. After that, there were 7557 unacceptable and 20114 acceptable ECG segments available.

ECG segments in COMD were used to train a conditional generative adversarial networks (CGANs) model for ECG segment generation at first. Then, generated unacceptable and real acceptable ECG data were used to pre-train a quality assessment model. Finally, training sets in COMD and RECD were both used for fine-tuning the assessment model, and testing sets were used to test the model. The detailed usage of data is illustrated in Table 2.

Table 1. The noise-add rules of dataset recreation.

Noise Type	MIT-BIHA	MIT-BIHNSR
bw	-	"19093", "19140", "19830"
em	-	All 17 recordings
ma	"101_V1", "106_V1", "112_V1", "113_V1", "114_V5", "115_V1", "122_V1", "200_V1", "205_V1", "209_V1", "215_V1", "220_V1", "221_V1", "222_MLII"	All 17 recordings
bw, g	-	Recordings expect "19093", "19140", "19830" (Total 14 recordings)
bw, p	-	Recordings expect "19093", "19140", "19830" (Total 14 recordings)
em, g	"101_V1", "106_V1", "112_V1", "113_V1", "114_V5", "115_V1", "122_V1", "200_V1", "205_V1", "209_V1", "215_V1", "220_V1", "221_V1", "222_MLII"	-
ma, bw	"112", "113", "114", "115", "116", "117", "118", "119", "121", "122", "123"	-
ma, em	"124", "200", "201", "202", "203", "205", "207", "208", "209", "210", "213", "214", "215"	All 17 recordings
bw, g, p	"101_V1", "106_V1", "112_V1", "113_V1", "114_V5", "115_V1", "122_V1", "200_V1", "205_V1", "209_V1", "215_V1", "220_V1", "221_V1", "222_MLII"	-
em, bw, g	"212", "217", "219", "220", "221", "228", "230", "231", "232", "233", "234"	-
ma, em, bw	"100", "101", "102", "103", "104", "105", "106", "107", "108", "109", "111"	-
g	first 5 min of each recording	first 5 min of each recording
p	first 5 min of each recording	first 5 min of each recording

Table 2. Usage of datasets.

Usage	COMD		Generated Unacceptable ECG	Parts of Acceptable ECG	RECD	
	Training Set	Testing Set			Training Set	Testing Set
Train CGANs	✓		-		-	
Pretrain Assessment Model	-		✓	✓		-
Finetune Assessment Model	✓	-	-	-	✓	-
Test Assessment Model	-	✓	-	-	-	✓

2.2. Methods

2.2.1. Data Augmentation

Insufficient and imbalanced data may reduce the performance of deep learning models [24,25]. Thus, the first procedure of the work was to automatically generate unacceptable ECG segments to solve data imbalance issues and perform data augmentation. Although traditional mathematical modeling methods can generate realistic heartbeat, the synthetic heartbeat's morphology lacks diversity or is even almost the same as those of training data [26]. Recently, several studies have confirmed that the generative adversarial networks (GANs) model has the ability to generate real-like ECG segments and arrhythmia [26–31]. The proposed system is shown in Figure 1, a GANs model [32] developed and trained based on COMD to obtain an ECG generator (G'), and G' is used to generate unacceptable ECG segments with 10 s duration (G'(z | y = 1)). The generator and discriminator are abbreviated as G and D, respectively. Figure 2(a) shows the structure of the proposed CGANs model. The label information is used as the condition, and each label ("0" for acceptable and "1" for unacceptable) is assembled to an M1-element vector representation, one input of G. The other input of G is a random M2-element noise signal and following the uniform distribution, their amplitude is limited in –1 to 1. Here, M1 and M2 are determined to be 20 and 700, respectively, by trials and errors. G mainly consists of two LSTM layers with 200 and 600 units, respectively. The main layers of D are two convolutional neural network (CNN) layers, with 128 and 64 units, respectively. Their kernel sizes are set to 10 and 5, respectively, and the alpha of the "LeakyReLU" activation layer is set to 0.2. The dropout rate is 0.3. The Dense layer has 32 units and uses "ReLu" as the activation function. For the output layer, we use "sigmoid" as its activation function. The Adam optimizer with a 0.0002 learning rate and binary crossentropy loss function are applied to train the CGANs model. By trials and errors, D is updated three times, and G is then updated once to train the model. After that, an ECG generator (G') is obtained from the CGANs model for unacceptable ECG segment generation.

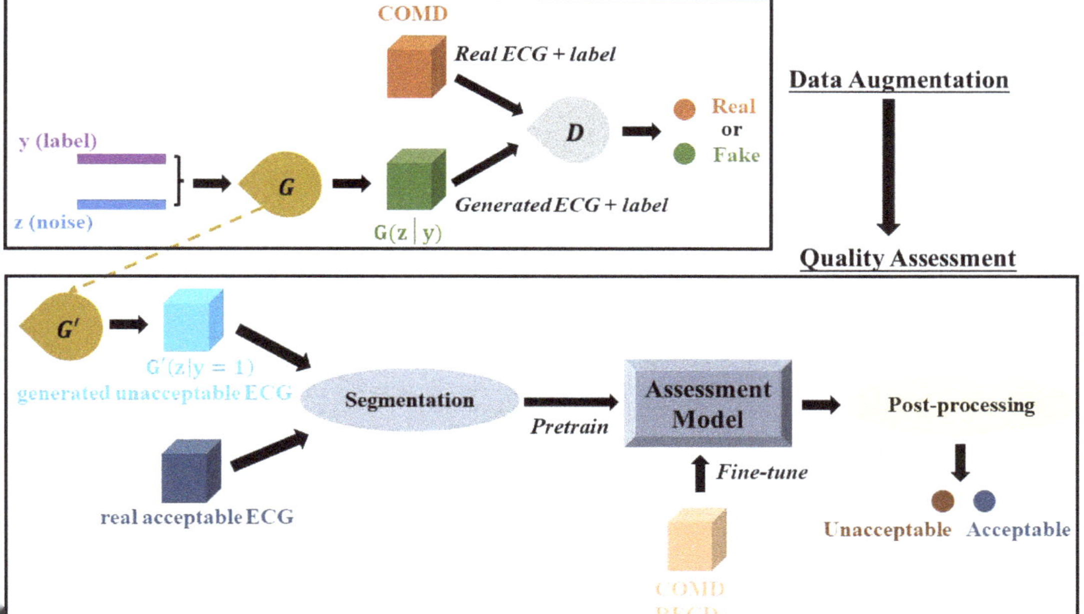

Figure 1. The proposed ECG quality assessment system. It consists of two parts: data augmentation by CGANs and a quality assessment model.

2.2.2. Quality Assessment

The generated unacceptable ECG segments and parts of real acceptable segments in RECD (5000 unacceptable and 5000 acceptable ECG segments in total) were used to pre-train the quality assessment model. The structure of the model is shown in Figure 2b, and consists of three branches: two CNN branches (branch1: left; branch2: middle) and an LSTM branch (branch3: right). For branch1, the number of filters in the two CNN layers is 128 and 32 with a kernel size of 50 and 10, respectively. A dropout rate was set to 0.3, and the pooling size to 10. Branch2 has the same structure as branch1, where its two CNN layers use 64 and 16 filters and the kernel size of each is 25 and 2, respectively. The dropout rate and pooling size are the same as branch1. The number of units in the two LSTM layers of branch3 are 200 and 100, respectively. The Dense layer has 32 units with a "ReLu" activation function. A batch size of 64, an Adam optimizer with a 0.0002 learning rate, and binary crossentropy loss function were applied for training. For model fine-tuning, the three branches were frozen and only parameters in the final two layers (Dense and Output layers) were updated with real data.

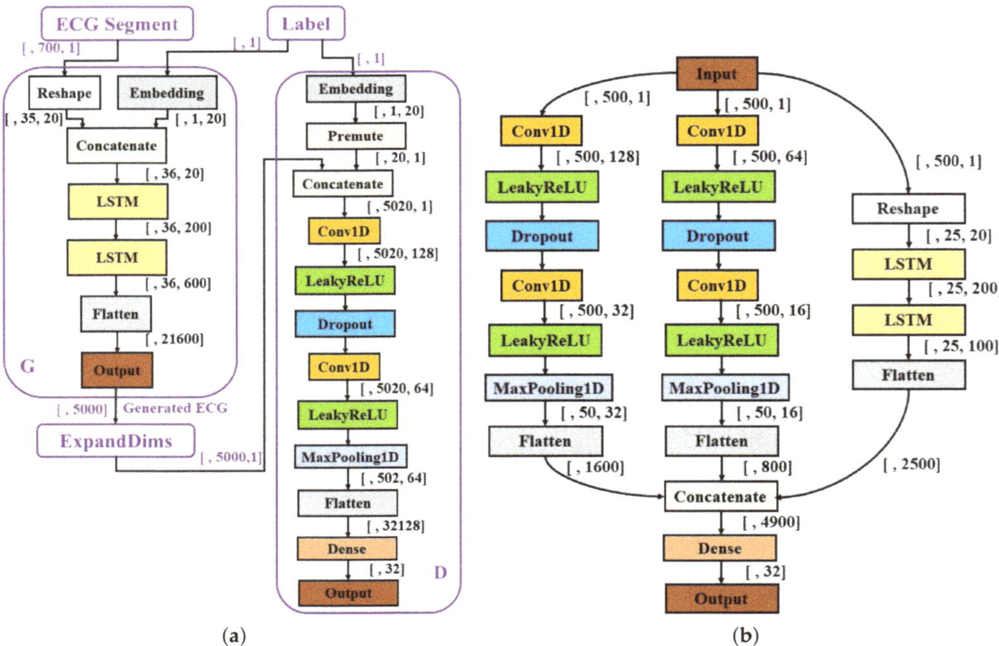

Figure 2. Structures of the proposed models. (**a**) The structure of CGANs; (**b**) the structure of the assessment model.

Considering the limited number of data in COMD, each 10 s ECG segment was further segmented into 10 examples with 1 s duration to increase the number of input examples for the training of the assessment model. Because our aim was to assess the quality of each 10 s ECG segment, we conducted a post-processing procedure as shown in Figure 3. The threshold was set to 3 for the quality assessment of ECG segment by trial an error. After model pre-training, the COMD and RECD datasets were randomly split 10 times to obtain training sets, which were used for model fine-tuning, and testing sets, which were used for testing the average performance of the model. For fair comparison, the same number of ECG segments in COMD and RECD datasets was used for fine-tuning (3000 segments with 1000 unacceptable and 2000 acceptable)), and the rest were used for testing.

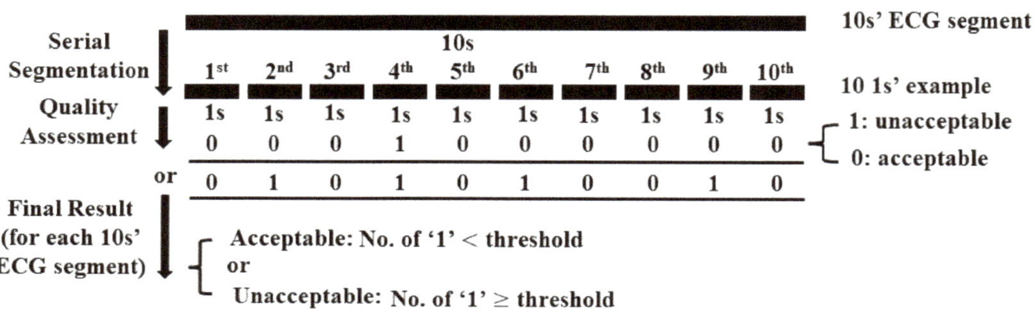

Figure 3. Post-processing for quality assessment of ECG segment.

3. Results

3.1. Data Augmentation

Figure 4 illustrates the training curves of CGANs and confirms their convergence. In Figure 4a, for the "Loss" figure, the blue line "d_real" means the loss of D when it is updated by real ECG segments, the orange line "d_fake" is the loss of D when it is updated by generated segments, and the green line "g" represents the loss of G. T-distributed stochastic neighbor embedding (tSNE) [33] was used to map real and generated segments to a three-dimensional space for visualization, as shown in Figure 4b. tSNE maps the similar segments to close points, and dissimilar segments were mapped to distant points. Figure 5 visualizes several real segments and generated segments by G'.

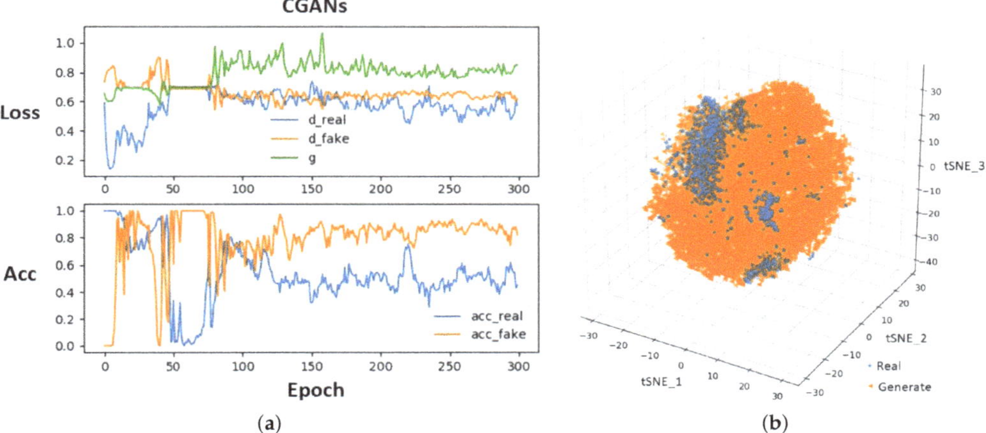

(a)

(b)

Figure 4. Training process of CGANs (**a**) and visualization of ECG segment distribution (**b**).

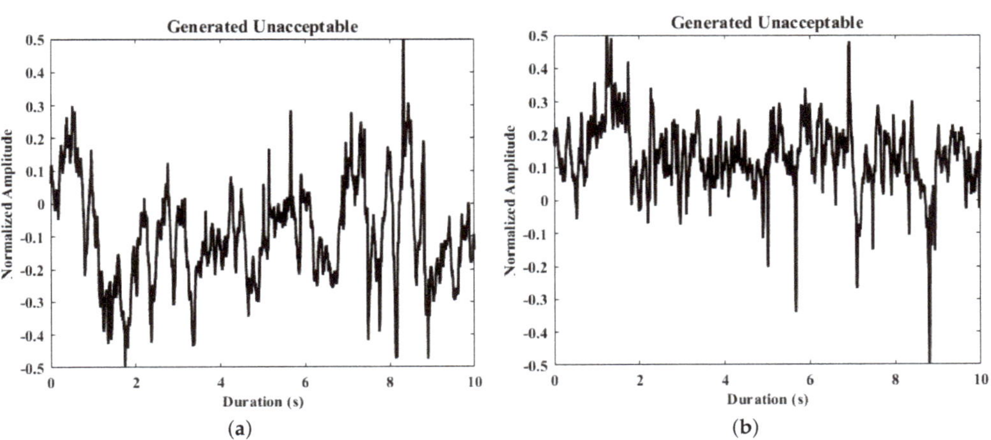

(a)

(b)

Figure 5. *Cont.*

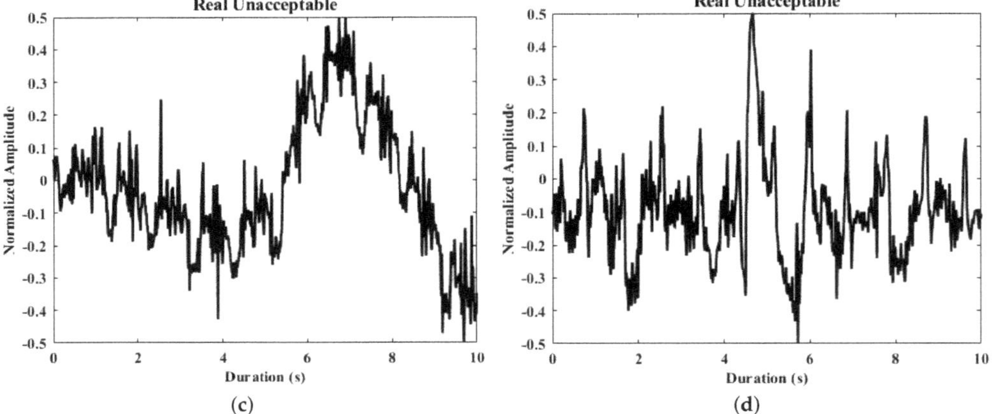

Figure 5. Samples of real and generated ECG segments. (**a,b**) The generated unacceptable ECG segments; (**c,d**) the real unacceptable ECG segments from the derivation dataset.

3.2. Quality Assessment

The performance of the proposed quality assessment model was measured by three indexes: accuracy, sensitivity, and specificity, which were calculated based on true positive (*TP*), true negative (*TN*), false positive (*FP*), and false negative (*FN*), as shown as follows:

$$Accuracy = \frac{TP + TN}{TP + TN + FP + FN},\tag{1}$$

$$Sensitivity = \frac{TP}{TP + FN},\tag{2}$$

$$Specificity = \frac{TN}{TN + FP}.\tag{3}$$

Figure 6a,b shows the average performance of the assessment model. The specificities are 96.4% and 95.0%, sensitivities are 98.6% and 99.1%, and accuracies are 97.1% and 96.4% respectively for COMD and RECD.

For comparison, some additional experiments were conducted as listed in Table A CGANs model for 1 s example generation was developed. The quality assessment model performs 95.5% accuracy (acc), 94.5% sensitivity (sen), and 96.0% specificity (spe) on COMD, all lower than the performance of the proposed system. In addition to CGANs, GANs model was also developed for data augmentation. The model failed in convergence when trained for 10 s ECG segments generation, but it was possible to generate 1 s ECG examples. Using the data generated by the GANs model, the quality assessment model shows an accuracy of 95.4%, sensitivity of 99.3% and specificity of 93.4% on COMD, which are not as good as the performance of the proposed method.

Moreover, to prove the necessity of training the quality assessment model using examples with a duration of 1 s, the quality assessment model was pre-trained directly using ECG segments with 10 s duration, and adding L2 regularization in CNN, LSTM and Dense layers to alleviate overfitting; however, the model still performed differently between the training and testing sets. For COMD, the training accuracy of the model was 95.8%, the sensitivity was 91.2%, and the specificity was 98.0%, while the testing accuracy, sensitivity, and specificity were 84.4%, 75.8%, and 88.2%, respectively. This may indicate that training by segments with 1 s may improve the generalization of the quality assessment model.

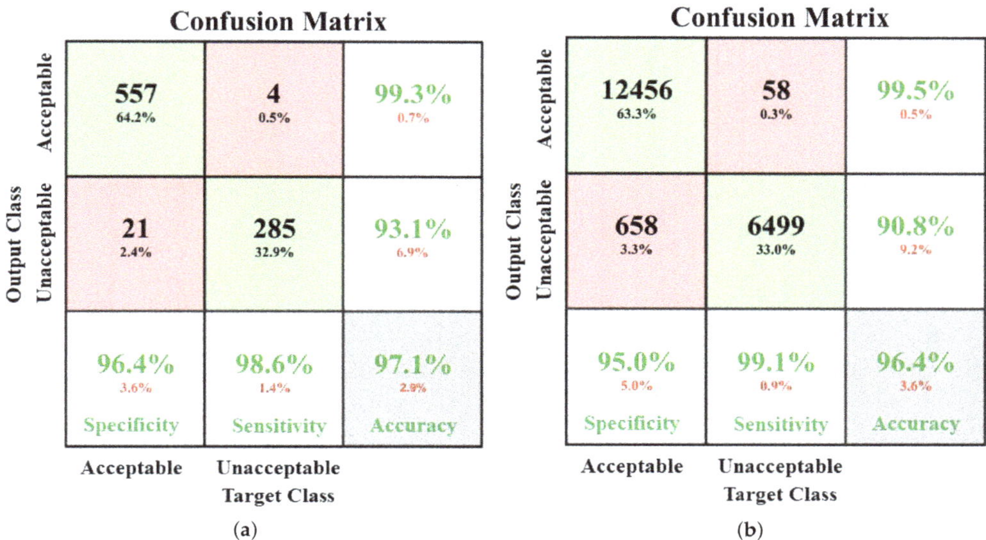

Figure 6. Performances of the proposed quality assessment model on COMD (**a**) and RECD (**b**).

Without data augmentation and pre-training the assessment model by generated ECG data, each 10 s ECG segment was segmented to 10 examples with 1 s duration to directly increase the number of examples and then use them to directly develop the assessment model. Its performance declined on both COMD and RECD compared with the proposed system, as listed in Table 3.

In addition to data generation, downsampling is a possible way to assist to train a deep learning model [34]. Thus, a comparison with a previous quality assessment method [35] was conducted. The previous study downsampled ECG segments from 500 Hz to 50 Hz and developed a shallow neural networks model to avoid overfitting. The accuracy of the proposed system increased by 1.3% and 2.6%, respectively, on COMD and RECD.

Table 3. Performance of models for quality assessment.

	Data Augmentation	Performance of Quality Assessment		Remark
Model	Duration of Generated ECG	COMD	RECD	
CGANs	10 s	acc: 97.1%; sen: 98.6%; spe: 96.4%	acc: 96.4%; sen: 99.1%; spe: 95.0%	Proposed method
CGANs	1 s	acc: 95.5%; sen: 94.5%; spe: 96.0%	-	-
GANs	10 s	-	-	GANs: convergence failed
GANs	1 s	acc: 95.4%; sen: 99.3%; spe: 93.4%	-	-

Table 3. *Cont.*

| Data Augmentation | | Performance of Quality Assessment | | Remark |
Model	Duration of Generated ECG	COMD	RECD	
CGANs	10 s	acc: 84.1%; sen: 75.8%; spe: 88.2%	-	Directly using 10 s ECG segments for assessment model development, and adding L2 regularization in CNN, LSTM, and Dense layers, but the model still performs overfitting; acc: 95.8% vs. 84.1% (trainng set vs. testing set); sen: 91.2% vs. 75.8%; spe: 98.0% vs. 88.2%
-	-	acc: 94.1%; sen: 96.5%; spe: 92.9%	acc: 94.0%; sen: 98.1%; spe: 91.9%	Without data augmentation, but segment each 10 s ECG segment to 10 examples with 1 s duration to naturally increase the number of examples.
-	-	acc: 95.8%; sen: 96.5%; spe: 95.5%	acc: 93.8%; sen: 89.0%; spe: 96.2%	Using shallow model and downsampled ECG segments, which is similar to the previous work [35], to avoid overfitting.

4. Discussion

In this study, ECG generated by CGANs benefited the ECG quality assessment tas Although downsampling is a way to assist to train deep learning models with a sm training dataset, it may limit the complexity of deep neural network models and th reduce the final performance. In this work, with the same training dataset, after p training with generated ECG data, the developed deep neural networks model obtained better performance (on COMD, data augmentation by CGANs vs. downsampling: 97.1 vs. 95.8% for accuracy, 98.6% vs. 96.5% for sensitivity and 96.4% vs. 95.5% for specifici on RECD, data augmentation by CGANs vs. downsampling: 96.4% vs. 94.0% for accura 99.1% vs. 89.0% for sensitivity and 95.0% vs. 96.2% for specificity). Finally, it just needed retrain two layers in the whole deep model; in this way, the size of the required traini set may be greatly reduced compared with training from scratch. This indicates that t GANs technique may be effective to assist the training of deep neural network models ECG-related decision-making, such as arrhythmia detection [28].

Traditional mathematical modeling methods are limited to synthesize a normal re istic heartbeat or ECG signals generally with the same morphology [36,37]. To gener ECG signals with arrhythmia, the traditional method needs to manually control positi parameters of P, Q, R, S, or T waves [36] to enrich the morphology of heartbeats. On t contrary, the GANs method can instinctively generate ECG signals with a larger divers which better matches with real ones [26]. The ECG signal is made up of temporal sequenc and thus, generating longer durations of ECG segments is preferred. In this work, CGA demonstrated a reliable convergence when they were trained for generating a 10 s EC segment, but GANs failed to converge. It may be attributed to how the discrimina in CGANs is required to not only identify generated or real ECG segments, but to a provide a correct label to each real segment. This gives a stronger constraint for mo training than GANs.

In this study, the proposed system assessed the quality of each 10 s ECG segm by consecutively analyzing the quality of 10 examples, each 1 s in length. This impro the reliability of the system. For example, for a 10 s ECG segment, a threshold of 3 set in post-processing, as shown in Figure 3; that is, if there were more than 3 out of 10 consecutive 1 s examples which were determined as "unacceptable" by the mo the corresponding 10 s ECG segment was classified as "unacceptable". In this case, des the quality of the rest of the seven consecutive examples, the final result did not cha and the sensitivity was assured. This characteristic is confirmed in the results, as shown in Figure 6; that is, the sensitivity of the model validated by COMD and RECD is higher than both the specificity and accuracy.

This work has some limitations, as follows.

(1) The proposed system is suitable for an initial ECG quality assessment, without considering specific applications. The quality requirement may vary for different purposes. For better explanation, several ECG segments that were labeled as "noisy" in PCCC2017 are shown in Figure 7. For example, for a task of AF detection, one of the characteristics for the detection of AF is that there are irregular heartbeats and no regular P waves in ECG. However, P waves in Figure 7a,b (Figure 7a: recording "A07/A07983", time: 10:10:15–10:10:25; Figure 7b: recording "A01/A01938", time: 05:05:40–05:05:50) were hardly observed because of severe noise; thus, the quality of ECG segments in Figure 7a,b is unacceptable because they cannot be used for further AF detection using P waves. In contrast, the segment in Figure 7a can be used for HRV time-domain analysis since it has obvious R peaks (green circle), and thus, the heart rate can be accurately calculated. In this case, it should be regarded as acceptable. Moreover, it can identify premature ventricular contraction (PVC) rhythms in Figure 7b (purple rectangle) and normal beats (green rectangles); thus, it is acceptable when used for PVC detection. In Figure 7c (recording "A00/A00445", time: 09:09:09–09:09:19), the ECG segment can be partly regarded as acceptable (green rectangle) or unacceptable (two sides). The ECG segment in Figure 7d (recording "A01/A01116", time: 04:02:47–04:02:57) should be totally unacceptable because the signal is completely contaminated by noise.

(2) In this study, the system only considered a quality assessment for single-lead ECG; therefore, the proposed method cannot be directly applied to 12-lead ECG quality assessments. The system is suitable for the quality assessment of ECGs collected by bedside monitors or wearable devices, but should be improved for the diagnosis of cardiovascular diseases using 12-lead ECG, such as acute myocardial infarction.

(3) In this study, the proposed system was unable to provide specific signal-to-noise ratio information for acceptable and unacceptable ECG signals; thus, it may be hard to quantize the quality of ECG signals.

In future, it is expected that we develop a multi-hierarchical and meticulous ECG quality assessment system. The system will identify low-quality ECG signals and perform task-specific quality assessments.

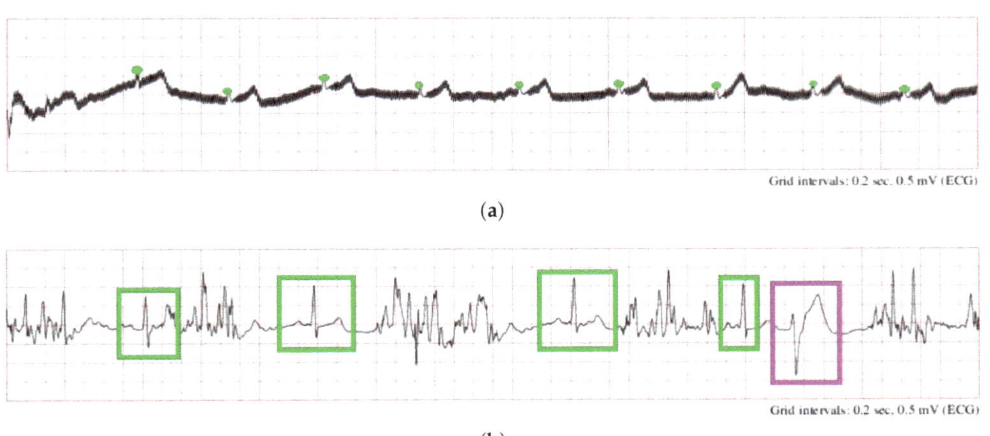

Figure 7. *Cont.*

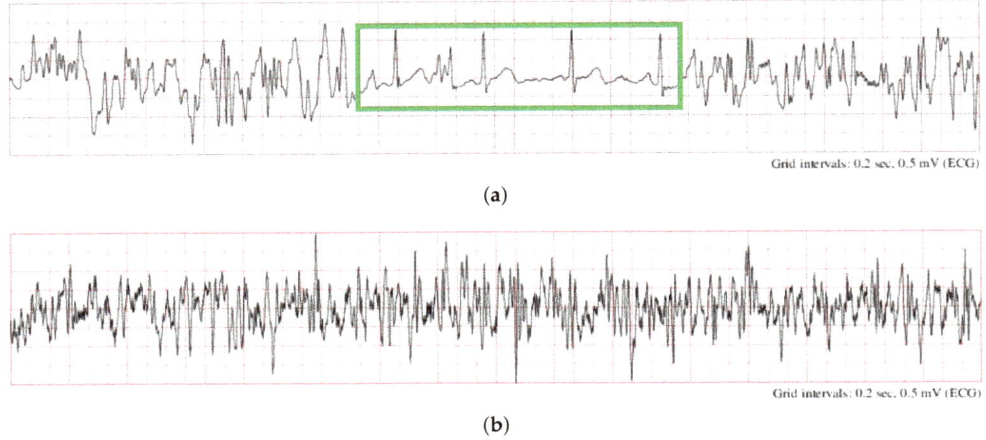

(a)

(b)

Figure 7. Samples of noisy ECG signals in PCCC2017 (**a**): recording "A07/A07983", time: 10:10:15–10:10:25; (**b**): recording "A01/A01938", time: 05:05:40–05:05:50; (**c**) recording "A00/A00445", time: 09:09:09–09:09:19, (**d**) recording "A01/A01116", time: 04:02:47–04:02:57 [18].

5. Conclusions

The CGANs technique is a possible method for ECG generation, and the generated data will help to improve the results of ECG quality assessments. The proposed system is expected to be applied for the accurate initial screening of ECG quality. In particular for patients with wearable ECG recording devices, the system may assist inexperienced users to collect ECG signals with a quality that meets diagnostic requirements. For clinical technicians, the proposed system is expected to relieve them from tedious and time consuming quality screening work.

Author Contributions: Conceptualization, X.Z. (Xue Zhou) and X.Z. (Xin Zhu); methodology, X. (Xue Zhou) and X.Z. (Xin Zhu); software, X.Z. (Xue Zhou); validation, X.Z. (Xue Zhou), X.Z. (Xin Zhu), K.N. and M.N.; formal analysis, X.Z. (Xue Zhou); investigation, X.Z. (Xue Zhou); resources, X. (Xin Zhu), K.N. and M.N.; data curation, X.Z. (Xue Zhou), X.Z. (Xin Zhu), K.N. and M.N.; writing—original draft preparation, X.Z. (Xue Zhou) and X.Z. (Xin Zhu); writing—review and editing, X. (Xue Zhou), X.Z. (Xin Zhu), K.N. and M.N.; visualization, X.Z. (Xue Zhou) and X.Z. (Xin Zhu; supervision, X.Z. (Xin Zhu); project administration, X.Z. (Xin Zhu); funding acquisition, X.Z. (X. Zhu) and K.N.. All authors have read and agreed to the published version of the manuscript.

Funding: This research was funded by JSPS Kakenhi Basic Research Fund C 18K11532 (X.Z.) and 21K10287 (K.N.), and Competitive Research Fund from The University of Aizu, 2021-P-5.

Institutional Review Board Statement: Not applicable.

Informed Consent Statement: Patient consent was waived due to the use of public databases.

Data Availability Statement: (1) The PhysioNet Computing in Cardiology Challenge 2017: http://physionet.org/content/challenge-2017/1.0.0/ (accessed on 1 December 2020); (2) TELE ECG Database: https://dataverse.harvard.edu/dataset.xhtml?persistentId=doi:10.7910/DVN/QTG0EP (accessed on 1 December 2020); (3) MIT-BIH arrhythmia database: https://physionet.org/content/mitdb/1.0.0/ (accessed on 1 December 2020); (4) MIT-BIH Normal Sinus Rhythm Database: https://www.physionet.org/content/nsrdb/1.0 (accessed on 1 December 2020).

Acknowledgments: TensorFlow 2.3.0 are freely provided from tensorflow.org to conduct experiments.

Conflicts of Interest: The authors declare no conflict of interest.

References

1. Baig, M.M.; Gholamhosseini, H.; Connolly, M.J. A comprehensive survey of wearable and wireless ECG monitoring systems for older adults. *Med. Biol. Eng. Comput.* **2013**, *51*, 485–495. [CrossRef] [PubMed]
2. Rosiek, A.; Leksowski, K. The risk factors and prevention of cardiovascular disease: The importance of electrocardiogram in the diagnosis and treatment of acute coronary syndrome. *Ther. Clin. Risk Manag.* **2016**, *12*, 1223–1229. [CrossRef] [PubMed]
3. Shaffer, F.; Ginsberg, J.P. An overview of heart rate variability metrics and norms. *Front. Public Health* **2017**, *5*, 285. [CrossRef] [PubMed]
4. Maršánová, L.; Ronzhina, M.; Smíšek, R.; Vítek, M.; Němcová, A.; Smital, L.; Nováková, M. ECG features and methods for automatic classification of ventricular premature and ischemic heartbeats: A comprehensive experimental study. *Sci. Rep.* **2017**, *7*, 1–11.
5. Satija, U.; Ramkumar, B.; Manikandan, M.S. A review of signal processing techniques for electrocardiogram signal quality assessment. *IEEE Rev. Biomed. Eng.* **2018**, *11*, 36–52. [CrossRef] [PubMed]
6. Improving the Quality of ECGs Collected Using Mobile Phones—The PhysioNet Computing in Cardiology Challenge 2011. Available online: https://www.physionet.org/content/challenge-2011/1.0.0/ (accessed on 3 August 2021).
7. Moody, G.B.; Muldrow, W.; Mark, R.G. A noise stress test for arrhythmia detectors. *Comput. Cardiol.* **1984**, *11*, 381–384.
8. Tsien, C.L.; Fackler, J.C. Poor prognosis for existing monitors in the intensive care unit. *Crit. Care Med.* **1997**, *25*, 614–619. [CrossRef]
9. Goldberger, A.L.; Amaral, L.A.; Glass, L.; Hausdorff, J.M.; Ivanov, P.C.; Mark, R.G.; Mietus, J.E.; Moody, G.B.; Peng, C.K.; Stanley, H.E. PhysioBank, PhysioToolkit, and PhysioNet: Components of a new research resource for complex physiologic signals. *Circulation.* **2000**, *101*, e215–e220. [CrossRef]
10. Quesnel, P.X.; Chan, A.D.; Yang, H. Real-time biosignal quality analysis of ambulatory ECG for detection of myocardial ischemia. In Proceedings of the 2013 IEEE International Symposium on Medical Measurements and Applications (MeMeA), Gatineau, QC, Canada, 4–5 May 2013.
11. Redmond, S.J.; Xie, Y.; Chang, D.; Basilakis, J.; Lovell, N.H. Electrocardiogram signal quality measures for unsupervised telehealth environments. *Physiol. Meas.* **2012**, *33*, 1517–1533. [CrossRef]
12. Shahriari, Y.; Fidler, R.; Pelter, M.M.; Bai, Y.; Villaroman, A.; Hu, X. Electrocardiogram signal quality assessment based on structural image similarity metric. *IEEE. Trans. Biomed. Eng.* **2017**, *65*, 745–753. [CrossRef]
13. Zhao, Z.; Zhang, Y. SQI quality evaluation mechanism of single-lead ECG signal based on simple heuristic fusion and fuzzy comprehensive evaluation. *Front. Physiol.* **2018**, *9*, 727. [CrossRef] [PubMed]
14. Moeyersons, J.; Smets, E.; Morales, J.; Villa, A.; De Raedt, W.; Testelmans, D.; Buyse, B.; Van Hoof, C.; Willems, R.; Van Huffel, S.; Varon, C. Artefact detection and quality assessment of ambulatory ECG signals. *Comput. Methods Programs Biomed.* **2019**, *182*, 105050. [CrossRef]
15. Fu, F.; Xiang, W.; An, Y.; Liu, B.; Chen, X.; Zhu, S.; Li, J. Comparison of Machine Learning Algorithms for the Quality Assessment of Wearable ECG Signals Via Lenovo H3 Devices. *J. Med. Biol. Eng.* **2021**, *41*, 231–240. [CrossRef]
16. MATLAB R2019b. Available online: https://www.tensorflow.org/learn (accessed on 11 December 2020).
17. TensorFlow. Available online: https://jp.mathworks.com/products/new_products/release2019b.html (accessed on 7 January 2021).
18. AF Classification from a Short Single Lead ECG Recording—The PhysioNet Computing in Cardiology Challenge 2017. Available online: https://physionet.org/content/challenge-2017/1.0.0/ (accessed on 1 December 2020).
19. Khamis, H.; Weiss, R.; Xie, Y.; Chang, C.W.; Lovell, N.H.; Redmond, S.J. TELE ECG Database: 250 Telehealth ECG Records (Collected Using Dry Metal Electrodes) with Annotated QRS and Artifact Masks, and MATLAB Code for the UNSW Artifact Detection and UNSW QRS Detection Algorithms. Harvard Dataverse, V3. Available online: https://dataverse.harvard.edu/dataset.xhtml?persistentId=doi:10.7910/DVN/QTG0EP (accessed on 1 December 2020).
20. Moody, G.B.; Mark, R.G. The impact of the MIT-BIH arrhythmia database. *IEEE Eng. Med. Biol. Mag.* **2001**, *20*, 45–50. [CrossRef] [PubMed]
21. MIT-BIH Normal Sinus Rhythm Database. Available online: https://physionet.org/content/nsrdb/1.0.0/ (accessed on 1 December 2020).
22. Khamis, H.; Weiss, R.; Xie, Y.; Chang, C.W.; Lovell, N.H.; Redmond, S.J. QRS detection algorithm for telehealth electrocardiogram recordings. *IEEE. Trans. Biomed. Eng.* **2016**, *63*, 1377–1388. [CrossRef] [PubMed]
23. The WFDB Software Package. Available online: https://archive.physionet.org/physiotools/wfdb.shtml (accessed on 1 December 2020).
24. Dong, Q.; Gong, S.; Zhu, X. Imbalanced deep learning by minority class incremental rectification. *IEEE Trans. Pattern Anal. Mach. Intelligence.* **2018**, *41*, 1367–1381. [CrossRef]
25. Japkowicz, N.; Stephen, S. The class imbalance problem: A systematic study. *Intell. Data Anal.* **2002**, *6*, 429–449. [CrossRef]
26. Wulan, N.; Wang, W.; Sun, K.; Xia, Y.; Zhang, H. Generating electrocardiogram signals by deep learning. *Neurocomputing* **2020**, *404*, 122–136. [CrossRef]
27. Goodfellow, I.J.; Pouget-Abadie, J.; Mirza, M.; Xu, B.; Warde-Farley, D.; Ozair, S.; Courville, A.; Bengio, Y. Generative adversarial nets. *Adv. Neural Inf. Process. Syst.* **2014**, *2014*, 2672–2680.
28. Zhu, F.; Ye, F.; Fu, Y.; Liu, Q.; Shen, B. Electrocardiogram generation with a bidirectional LSTM-CNN generative adversarial network. *Sci. Rep.* **2019**, *9*, 6734. [CrossRef]

29. Wang, P.; Hou, B.; Shao, S.; Yan, R. ECG arrhythmias detection using auxiliary classifier generative adversarial network and residual network. *IEEE Access* **2019**, *7*, 100910–100922. [CrossRef]
30. Ye, F.; Zhu, F.; Fu, Y.; Shen, B. ECG generation with sequence generative adversarial nets optimized by policy gradient. *IEEE Access* **2019**, *7*, 159369–159378. [CrossRef]
31. Hazra, D.; Byun, Y. SynSigGAN: Generative Adversarial Networks for Synthetic Biomedical Signal Generation. *Biology* **2020**, *9*, 441. [CrossRef] [PubMed]
32. Mirza, M.; Osindero, S. Conditional generative adversarial nets. *arXiv* **2014**, arXiv:1411.1784.
33. Maaten, L.V.D.; Hinton, G. Visualizing data using t-SNE. *J Mach Learn Res.* **2008**, *9*, 2579–2605.
34. Jeon, E.; Oh, K.; Kwon, S.; Son, H.; Yun, Y.; Jung, E. S.; Kim, M. S. A Lightweight Deep Learning Model for Fast Electrocardiographic Beats Classification With a Wearable Cardiac Monitor: Development and Validation Study. *JMIR Med. Inform.* **2020**, *8*, e17037. [CrossRef] [PubMed]
35. Zhou, X.; Zhu, X.; Nakamura, K.; Noro, M. ECG quality assessment using 1D-convolutional neural network. In Proceedings of the 2018 14th IEEE International Conference on Signal Processing (ICSP), Beijing, China, 12–16 August 2018; pp. 780–784.
36. McSharry, P. E.; Clifford, G. D.; Tarassenko, L.; Smith, L. A. A dynamical model for generating synthetic electrocardiogram signals. *IEEE Trans. Biomed. Eng.* **2003**, *50*, 289–294. [CrossRef] [PubMed]
37. Roonizi, E. K.; Sameni, R. Morphological modeling of cardiac signals based on signal decomposition. *Comput. Biol. Med.* **2013**, *4*, 1453–1461. [CrossRef]

 *life*

Article

Identification of Targeted Proteins by Jamu Formulas for Different Efficacies Using Machine Learning Approach

Sony Hartono Wijaya [1,2,*], Farit Mochamad Afendi [2,3], Irmanida Batubara [2,4], Ming Huang [5], Naoaki Ono [5], Shigehiko Kanaya [5] and Md. Altaf-Ul-Amin [5,*]

[1] Department of Computer Science, IPB University, Kampus IPB Dramaga Wing 20 Level 5, Bogor 16680, Indonesia
[2] Tropical Biopharmaca Research Center, IPB University, Kampus IPB Taman Kencana, Bogor 16128, Indonesia; fmafendi@apps.ipb.ac.id (F.M.A.); ime@apps.ipb.ac.id (I.B.)
[3] Department of Statistics, IPB University, Kampus IPB Dramaga Wing 22 Level 4, Bogor 16680, Indonesia
[4] Department of Chemistry, IPB University, Kampus IPB Dramaga Wing 1 Level 3, Bogor 16128, Indonesia
[5] Computational Systems Biology Laboratory, Graduate School of Science and Technology, Nara Institute of Science and Technology, 8916-5 Takayama, Ikoma 630-0192, Nara, Japan; alex-mhuang@is.naist.jp (M.H.); nono@is.naist.jp (N.O.); skanaya@gtc.naist.jp (S.K.)
* Correspondence: sony@apps.ipb.ac.id (S.H.W.); amin-m@is.naist.jp (M.A.-U.-A.)

Abstract: Background: We performed in silico prediction of the interactions between compounds of Jamu herbs and human proteins by utilizing data-intensive science and machine learning methods. Verifying the proteins that are targeted by compounds of natural herbs will be helpful to select natural herb-based drug candidates. Methods: Initially, data related to compounds, target proteins, and interactions between them were collected from open access databases. Compounds are represented by molecular fingerprints, whereas amino acid sequences are represented by numerical protein descriptors. Then, prediction models that predict the interactions between compounds and target proteins were constructed using support vector machine and random forest. Results: A random forest model constructed based on MACCS fingerprint and amino acid composition obtained the highest accuracy. We used the best model to predict target proteins for 94 important Jamu compounds and assessed the results by supporting evidence from published literature and other sources. There are 27 compounds that can be validated by professional doctors, and those compounds belong to seven efficacy groups. Conclusion: By comparing the efficacy of predicted compounds and the relations of the targeted proteins with diseases, we found that some compounds might be considered as drug candidates.

Keywords: compound–protein interaction; Jamu; machine learning; drug discovery; herbal medicine

Citation: Wijaya, S.H.; Afendi, F.M.; Batubara, I.; Huang, M.; Ono, N.; Kanaya, S.; Altaf-Ul-Amin, M. Identification of Targeted Proteins by Jamu Formulas for Different Efficacies Using Machine Learning Approach. *Life* **2021**, *11*, 866. https://doi.org/10.3390/life11080866

Academic Editor: Stefania Lamponi

Received: 30 June 2021
Accepted: 18 August 2021
Published: 23 August 2021

Publisher's Note: MDPI stays neutral with regard to jurisdictional claims in published maps and institutional affiliations.

1. Introduction

Identification of compounds derived from herbal medicines and natural products has shown potential in drug discovery and drug development [1,2]. Many useful compounds have been found and utilized from herbal medicines and natural products to treat various diseases, such as oseltamivir [3] and roscovitine [4]. Oseltamivir is a neuraminidase inhibitor used in the treatment and prophylaxis of both influenza A and influenza B, whereas roscovitine is known as an anticancer drug. However, the process of identification of compound and target protein interactions in vivo and in vitro requires enormous effort. Therefore, efficient in silico screening methods are needed to predict the interaction between compounds and target proteins. In this light, in silico prediction of the interactions between compounds and target proteins can help in making the efforts easier.

As a country with the largest medicinal plant species in the world, Indonesians utilize medicinal plants as a constituent of herbal medicines [5–7]. These are known as Indonesian Jamu. Currently, Jamu is produced commercially on an industrial scale under the supervision of the National Agency of Drug and Food Control (NADFC) of Indonesia. Jamu, like the other herbal medicine systems, i.e., traditional Chinese medicine, Japanese Kampo, Ayurveda, and Unani, can be considered as a new resource for compound screening. The molecules might be from a specific part of a plant used as a Jamu ingredient, such as rhizome of Java ginger (*Curcuma xanthorrhiza*), leaf of *kecibeling* (*Strobilanthes crispus*), or fruit of tamarind (*Tamarindus indica*). The utilization of herbal medicines in drug screening is very promising because of the lack of side effects [8,9]. In addition, the high biodiversity in Indonesia has great advantages in the process of finding potential compounds in Jamu. Furthermore, the systematization of Jamu medicine might help not only to obtain information about the major ingredient plants in Jamu medicines, but also to find compound and protein interactions to explain formulation of Jamu. The information on interactions between Jamu compounds and human target proteins will allow understanding the mechanisms of how Jamu medicines work against diseases and will be helpful for finding new drugs based on a scientific basis.

Various screening approaches have been developed to determine candidate compounds from herbal medicines and natural products in drug discovery. One category of the popular approaches is machine learning techniques. This approach can learn from the data and the resulting model can be applied to make a prediction. Support vector machine (SVM) and random forest are machine learning methods for supervised learning, and they have been used in many research fields with success [10–12]. In order to obtain a good model, the machine learning method requires a great number of data samples. Nowadays, there are many open access databases that can be used to support the prediction of compound and protein interactions, such as KEGG [13], DrugBank [14], KNApSAcK [15], UniProt [16], and Online Mendelian Inheritance in Man (OMIM) [17]. Prediction of compound–protein interactions can exploit these databases to identify candidate compounds. In terms of Indonesian Jamu, IJAH Analytics can be considered as a good reference for Jamu because it has information about plant species used in Jamu formulas. In addition, plant species information can be associated with information regarding compounds, target proteins, diseases, and interactions between entities. It is hoped that the more efficient and effective application of natural products will improve the drug discovery process.

Many studies on the prediction of interactions between compounds and target proteins have been reported. Yamanishi et al. implemented a systematic study on the prediction of compound–target protein interactions by utilizing supervised learning using a bipartite graph [18]. The interactions were predicted by utilizing the structural similarity of compounds and the similarity of amino acid sequences. They computed the structural similarities between compounds using SIMCOMP and the sequence similarities between proteins using normalized Smith–Waterman scores [19,20]. In the prediction methods, they applied the bipartite local model (BLM) and SVM to predict compound–target protein interactions [21,22]. BLM predicts target proteins of a given compound using the structural similarity of compounds, proteomic similarity, and information of interactions between compounds and target proteins, whereas SVM was used as the classifier for the BLM.

In this study, we applied machine learning techniques to predict the interaction between compound and protein. SVM and random forest have been chosen as classifiers and compound and protein are represented by fingerprint and numerical representation of amino acid, respectively. The accuracy, sensitivity, and specificity were used in the evaluation of the models. After we confirmed the best model obtained in the prediction compound–protein interactions, we determine targeted proteins for candidate compound obtained from plants used in the Jamu formulas for different efficacies [11]. The objective was not only to identify targeted proteins for developing new drugs, but also to give comprehensive understanding of Jamu medicines on the molecular level.

2. Materials and Methods

Jamu medicines consist of a combination of medicinal plants and are used to treat various diseases. In this work, we exploit information about compound and protein interactions from open access databases to predict compound–protein interactions in the context of Jamu formulas. The concept of the proposed method is depicted in Figure 1, which mainly consists of three processes: (a) data transformation, (b) model generation and evaluation, and (c) prediction of targeted proteins by Jamu formulas.

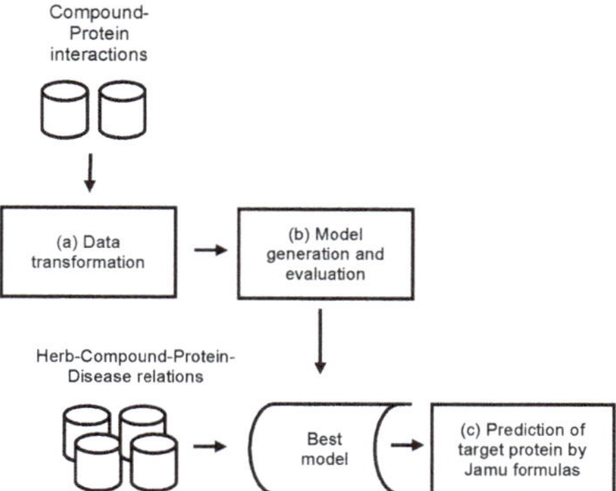

Figure 1. Concept of the methodology.

Initially, we collected the required data for this study from open access databases such as DrugBank, PubChem [23], KNApSAcK, UniProt, KEGG, OMIM, Matador [24], and Indonesian Jamu Herbs (IJAH Analytics, http://ijah.apps.cs.ipb.ac.id, accessed on 20 August 2021). The acquisition of data for generating the prediction model includes compounds, target proteins, and interactions between them. The chemical structures of the compounds were represented by Simplified Molecular Input Line Entry Specification (SMILES) codes. Many databases, such as DrugBank, provide SMILES of each compound [25]. We eliminated some compounds that have ambiguous SMILES or do not have SMILES information. Compounds with known SMILES codes were used in the training process to generate a model for predicting compound–protein interactions. In addition, the information about target proteins was also collected from public databases, especially the IJAH database, and these data were represented by amino acid sequences using FASTA format. In the case of interactions, we gathered that information from IJAH, Matador, and KEGG databases. We also collected the candidate compounds of Jamu formulas associated with efficacy groups from a previous study [11] and used those as test data.

2.1. Data Transformation

We transformed information about compounds and amino acid sequences into fingerprints and numerical representations of amino acids, respectively. In the case of compounds, we examined two different fingerprint representations, namely the binary representation of the Molecular Access System (MACCS) and PubChem fingerprints [12,26,27]. Therefore, each compound was represented as 166 and 881 binary vectors, respectively. In the case of proteins, we transformed amino acid sequences into the amino acid composition (AAC) and dipeptide composition descriptors [28]. The AAC represents an amino acid sequence as a fraction of each amino acid type within a protein, and it will produce 20-dimensional AAC vectors. The fractions of all 20 natural amino acids are calculated as:

$$f(r) = Nr/N, \ r = 1, 2, \ldots, 20 \tag{1}$$

where N_r is the number of the amino acid type r and N is the length of the sequence. In addition, dipeptide composition will produce 400-dimensional descriptors, defined as:

$$f(r,s) = \frac{N_{rs}}{N-1}, r, s = 1, 2, \ldots, 20 \tag{2}$$

where N_{rs} is the number of dipeptides represented by amino acid type r and type s.

After we transformed compounds and proteins into fingerprints and numerical de
scriptors, we created four datasets consisting of all combinations of compound and protein
vectors for generating the model as follows: a combination of MACCS fingerprint and
AACs (called dataset 1), a combination of MACCS fingerprint and dipeptide descripto
(called dataset 2), a combination of PubChem fingerprint and AACs (called dataset 3), and
a combination of PubChem fingerprint and dipeptide descriptor (called dataset 4). Figure 2
illustrates the data representation of compounds, proteins, and interactions between them
In the case of testing data, we built combinations of candidate compounds from medicina
plants in Jamu and proteins.

Samples	Features																			Class
	CD1	CD2	CD3	CD4	CD5	CD6	CD7	CD8	...	CDm	PD1	DP2	PD3	PD4	PD5	PD6	PD7	PD8	... PDn	Interaction
DT1																				1
DT2																				1
DT3																				0
DT4																				0
DT5			Molecular fingerprints										Numerical protein descriptors							0
DT6																				1
DT7																				0
...																				...
DTk																				0

Figure 2. Data representation. Each data sample DT_k is composed of molecular fingerprints (CD
CD_2, CD_3, ..., CD_m) and numerical protein descriptors (PD_1, PD_2, PD_3, ..., PD_n).

2.2. Model Generation and Evaluation

We applied SVM and random forest in the model generation step. SVM is a bina
classifier based on constructing an optimal linear model, which has the largest margi
between two classes. The linear separator is constructed by simultaneous minimization
the empirical classification error and maximization of the geometric margin [29]. If we hav
n training data pairs, $T = \{(x_i, y_i)\}$, $i = 1, \ldots, n$, where $x_i (\in \mathbb{R}^p)$ is a vector representin
compound and protein and y_i is the class of x_i. The decision function of SVM is defined
$f(x) = w^T x + b$, where $w = [w_1, w_2, \ldots, w_p]^T$ is the weight vector and b is a scalar. Th
optimization problem that SVMs aim to minimize is shown in Equation (3):

$$\min_{w \in \mathbb{R}^p, \xi_i \in \mathbb{R}^+} \frac{1}{2} \|w\|^2 + C \sum_i^n \xi_i$$

subject to $y_i(w^T x_i + b) \geq 1 - \xi_i$, where C is a trade-off between the width of the margin a
the number of misclassifications, and ξ_i is a slack variable. SVM can be extended to class
data that are not linearly separable by utilizing a kernel technique. There are two kern
functions that we applied in this study, namely the linear kernel $(K(x_i, x_j) = x_i^T, x_j)$ a
radial basis function (RBF) kernel $(K(x_i, x_j) = e^{-\gamma \|x_i - x_j\|^2}, \gamma > 0)$, where γ is the inver
of the radius of influence of samples selected by the model as support vectors [10,30].

In addition, random forest is an ensemble method composed of many decision tre
For each classification tree, a bootstrap sample of the data is generated, and at each sp
the candidate set of variables is a random subset of the variables [31–33]. Given a
of training samples $L = \{(x_i, y_i)\}, i = 1, \ldots, n$, where $x_i (\in \mathbb{R}^p)$ is a vector of predic
variables representing compound–protein data i and y_i is the class label. Random for
targets generating a number of *ntree* decision trees from these samples. The same num
of n samples is randomly selected with replacement (bootstrap resampling) for each tre
form a new training set, and the samples not selected are called out-of-bag (OOB) samp
Using this new training set, a decision tree is grown to the largest extent possible with
any pruning according to the classification and regression tree (CART) methodology [

The Gini index is used during the development process of a decision tree. The Gini index at node v, $Gini(v)$, is shown in Equation (4).

$$Gini(v) = \sum_{c=1}^{C} \hat{p}_c^v (1 - \hat{p}_c^v) \tag{4}$$

where $\hat{p}_c^v$ is the proportion of class c observations at node v [35]. Then, the Gini information gain of x_i for splitting node v into two child nodes, $Gain(x_i, v)$, is shown in Equation (5):

$$Gain(x_i, v) = Gini(x_i, v) - w_L Gini(x_i^L, v^L) - w_R Gini(x_i^R, v^R) \tag{5}$$

where v^L and v^R are the left and right child nodes of v, w_L and w_R are the proportions of instances assigned to the left and right child nodes, and x_i^L and x_i^R are the instances in the left and right child nodes. At each node, a random set of *mtry* features out of p is evaluated, and the feature with the maximum $Gain(x_i, v)$ is used for splitting the node v. The OOB error is estimated in the process of constructing the forest. After constructing the entire forest, OOB classification results for each sample are used to determine a decision for this sample via a majority-voting rule.

We defined and compared the performance of the resulting models by using accuracy, sensitivity, and specificity [36,37]. The higher the accuracy is, the better the performance of the classifier is. We measured the accuracies of SVM with two different kernels and random forest using four data representations (datasets 1–4). In order to estimate the performance of random forest and SVM with two different kernels, 10-fold cross-validation was used [21]. Each of the datasets was divided into 10 subsamples. Then, nine samples were used as a training dataset to make a classification model, and the remaining sample was used as a validation dataset for testing the model. In the model evaluation step, we selected the best classifier and data representation of compounds and amino acid sequences for which we obtained the best result and used that for the prediction of target proteins.

2.3. Prediction of the Target Protein by Jamu Formulas

The best model with the highest accuracy was applied for the prediction of compound–protein interactions concerning Jamu formulas used as the testing dataset. In this case, we accepted compound–protein interactions as true interactions when the probability was greater than a threshold. Figure 3 illustrates the relations among different entities involving comprehensive Jamu research, where a dotted rectangle indicates the focus of the present work. Figure 3 also shows how we validate our results by comparing efficacy–compound and protein–disease relations. We validated the results by comparing the therapeutic usage of predicted compounds and the relations of the targeted proteins with diseases. We assessed and discussed the results with supporting evidence from published literature and comments from professional doctors.

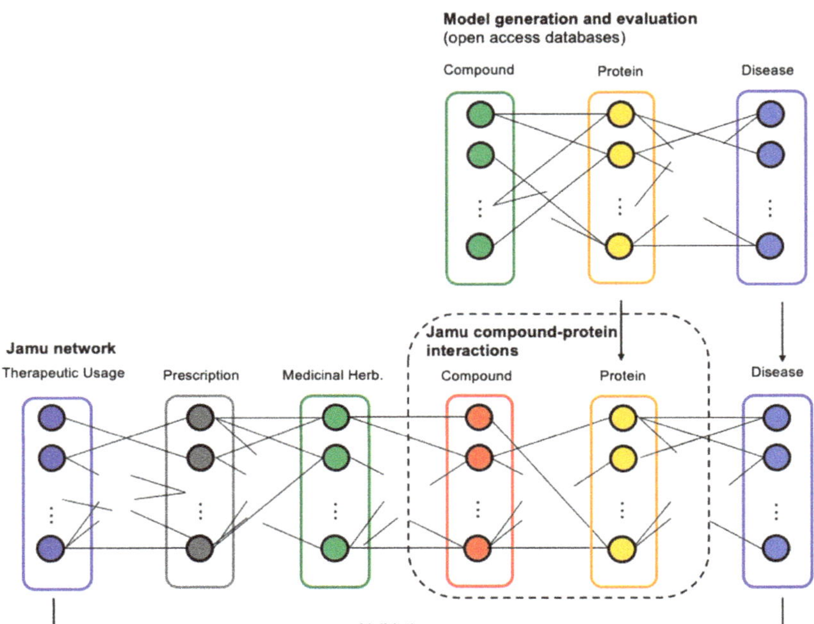

Figure 3. The process of prediction and identification of targeted proteins. Initially, we developed prediction model of compound–protein interactions by utilizing compound–protein data. Then, the best model was used to predict interaction between compounds of Jamu formulas and their targeted proteins (in the dotted rectangle).

3. Results and Discussion

The summary of data used in this study is shown in Table 1. We utilized compounds that are reported to be available in the herbs used as Jamu ingredients. There are 17.22 compounds belonging to 4.984 Indonesian herbs collected from KNApSAcK, IJAH, PubChem, and KEGG databases. In addition, the number of target proteins collected from UniProt and IJAH databases is 3.334, and the number of interactions collected from UniProt, IJAH, Matador, and KEGG databases is 7.989. Initially, we removed the data that do not have necessary properties, such as the SMILES in the case of the compounds and the amino sequence in the case of target proteins. Furthermore, we removed the compound and proteins that are not included in the compound–protein interactions data. We also considered only those compounds that target human proteins. Therefore, the number of compounds, proteins, and interactions used in this experiment are 2.146, 3.334, and 7.216, respectively.

3.1. Preprocessing of Compound and Protein Data

The transformation of compounds from SMILES to fingerprints was done by utilizing ChemDes web-based software and PaDEL descriptor [27,38]. Compounds were transformed to MACCS and PubChem fingerprints. Moreover, we used the protr package in R to generate AAC and dipeptide as numerical representation schemes of protein sequences [28]. We eliminated two amino acid sequences in the preliminary study, i.e., Q9NZV5 and P36969, because they showed unrecognized amino acid type when transforming amino acid sequences to AAC. Therefore, there were 3.332 proteins left for further processes.

After data transformation finished, we created datasets for compound–protein prediction using both compound and protein space information. Each sample vector is composed of a fingerprint and numerical descriptor of compound and protein. Therefore, for two different compound fingerprints and two protein numerical descriptors, we constructed four matrices with dimensions (2.146 × 3.332) by (166 + 20) for MACCS + AAC, (166 + 400) for MACCS + dipeptide, (881 + 20) for PubChem + AAC, and (881 + 400) for PubChem + dipeptide. The information of interactions between compounds and proteins is considered as a positive class, whereas unknown interactions are considered as a negative class. As the number of samples in the negative class is significantly large (number of compounds multiplies the number of proteins), we randomly selected 7.216 samples for the negative class, the same as the number of positive samples. We determined positive and negative class interactions as classes 1 and 0, respectively.

Wijaya et al. [11] identified 94 significant compounds associated with twelve efficacy groups, and 28 of them were validated by published literature. In this case, the efficacy refers to broad disease classes which are as follows: blood and lymph diseases (E1), cancers (E2), the digestive system (E3), female-specific diseases (E4), the heart and blood vessels (E5), male-specific diseases (E6), muscle and bone (E7), nutritional and metabolic diseases (E8), respiratory diseases (E9), skin and connective tissue (E10), the urinary system (E11), and mental and behavioral disorders (E12). We considered those 94 compounds as test data in this study. Table 2 shows the number of candidate compounds for each efficacy. We transformed the compounds into fingerprints according to the best results we obtained.

Table 1. The distribution of compound, protein, and interaction between them as training and testing data.

Description	Number of Data	Identifier	References
Protein	3.334	UniProtID	UniProt, IJAH
Compound	17.277	CAS_ID, PubChem ID, KEGG ID	KNApSAcK, PubChem, KEGG, IJAH
Compound of Jamu	94	Compound ID	Wijaya et al. [11]
Compound–protein interactions	149		KEGG
	4.144		Matador
	3.696		UniProt, IJAH
Amino acid sequences	3.334	UniProtID	UniProt

Table 2. The number of compounds for predicting target proteins. All data are classified by efficacies, and some compounds are related to one or more efficacy groups.

ID	Efficacy Groups	Number of Compounds
E1	Blood and Lymph Diseases	15
E2	Cancers	5
E3	The Digestive System	17
E4	Female-Specific Diseases	16
E5	The Heart and Blood Vessels	4
E6	Male-Specific Diseases	5
E7	Muscle and Bone	18
E8	Nutritional and Metabolic Diseases	7
E9	Respiratory Diseases	32
E10	Skin and Connective Tissue	4
E11	The Urinary System	14
E12	Mental and Behavioral Disorders	8

3.2. Prediction Performance

We applied the R packages named e1071 ver. 1.7–4 to implement the SVM method [39] and randomForest ver.4.6–14 to implement random forest (https://cran.r-project.org/web/packages/randomForest/, accessed on 9 August 2020). The optimal parameters used in the model generations were obtained by utilizing best.tune and tuneRF functions for SVM and random forest, respectively. In the SVM, the regulation parameter C depends on numerical protein descriptors. In the case of AACs, C is equal to 1, whereas C is equal to 1000 in dipeptide. The γ values of datasets 1–4 are 0.00763, 0.00177, 0.00437 and 0.00078, respectively. In random forest, the appropriate number of trees *ntree* for datasets 1 anc 3 is the same, 1000. Additionally, the *ntree* values for datasets 2 and 4 are 2000 and 50C respectively. The *mtry* values for dataset 2 and 4 are the same, i.e., 10, whereas those fo datasets 1 and 2 are 6 and 15, respectively.

Table 3 shows the prediction performance for each type of dataset and each mode Representation of amino acid sequences using AAC descriptor in datasets 1 and dataset ? obtains better accuracy compared to dipeptide descriptor on both classifiers and com pound fingerprints. Furthermore, if we compare the performance of random forest anc support vector machine classifiers, the classification accuracy of random forest using AAC descriptor is better than SVM with both kernels. In the case of fingerprints that are used t represent the compounds, MACCS obtains slightly better classification results than Pul Chem features. One of the reasons for the poor classification results on the dataset usin the dipeptide descriptor is the number of features produced by the method. Dipeptid makes 400 features, causing the number of compound–protein features representing th input data to increase. Many features have zero values and affect the resulting model. is very challenging to determine the most appropriate features because machine learnin methods generally rely on feature engineering [40]. This can also be observed in datasets and 4 between MACCS and PubChem fingerprints; when the number of features increase this also reduces the resulting accuracy. Since this represents sufficiently high perfc mance, the model can be applied to predict interactions between the Jamu compounds an target proteins.

Table 3. The evaluation of generated models.

Datasets	Classifiers	Accuracy		Sensitivity		Specificity	
MACCS_AAC	SVM Linear	69.16%±	1.07%	71.52%±	1.84%	66.77%±	1.87%
	SVM RBF	81.71%±	1.52%	82.79%±	2.27%	80.62%±	1.27%
	Random Forest	89.30%±	0.69%	87.86%±	1.20%	90.74%±	1.05%
MACCS_Dipeptide	SVM Linear	61.68%±	0.77%	63.49%±	1.61%	61.27%±	0.88%
	SVM RBF	72.71%±	0.86%	71.81%±	1.81%	73.15%±	1.15%
	Random Forest	60.79%±	1.20%	59.14%±	1.56%	61.17%±	1.30%
PubChem_AAC	SVM Linear	70.77%±	0.90%	73.08%±	1.86%	68.49%±	1.87%
	SVM RBF	80.01%±	1.35%	80.52%±	1.80%	79.51%±	1.82%
	Random Forest	89.28%±	0.40%	87.96%±	0.88%	90.63%±	0.58%
PubChem_Dipeptide	SVM Linear	50.49%±	1.08%	54.15%±	1.38%	50.47%±	1.01%
	SVM RBF	49.55%±	1.28%	54.83%±	5.44%	49.56%±	1.19%
	Random Forest	50.28%±	0.72%	50.12%±	1.60%	50.28%±	0.71%

3.3. Prediction Results

In order to predict interactions between compounds and target proteins, the cl sification model was taken from the models that obtained the best classification resu Additionally, a testing dataset was constructed to match the dataset that achieved t best classification result. In this case, we utilized MACCS fingerprint to represent Jar compounds, AAC descriptor to represent amino acid sequences, and random forest a classifier. Since we focused on whether compounds bind to target proteins, we create matrix containing all combinations of candidate compounds of Jamu formulas and tar; proteins as shown in Figure 2. Then, the prediction model was applied to predict whet compound and protein have an interaction or not. We accepted compound–protein inte tions as true interactions when their classification probability was greater than 0.85. N all candidate compounds identified in the work of Wijaya et al. have interactions w one or more proteins that were utilized in the current experiment. Here, we predict 168 compound–protein interactions of Jamu formulas, involving 68 candidate compour Moreover, the professional doctors validated the predicted compound–protein interacti by comparing the efficacy of predicted compounds and the relations of the targeted prote with diseases, as shown in Figure 3. Based on the current results, interactions involv 27 compounds can be validated, and those compounds belong to seven efficacy grou Table 4 summarizes predicted compound–protein interactions by Jamu formulas that h been validated by professional doctors, and all of them are presented under respective efficacies. We also discovered a protein is targeted by many compounds and a compound has interaction with many target proteins. For instance, P02768, known as human serum albumin (HSA), is targeted by caffeic acid, diacetoxy-6-gingerdiol, gallic acid, luteolin, quercitrin, tricin, and ursolic acid. In addition, ursolic acid targets Q92887, Q9NPD5, Q9Y6L6, P08185, and P02768. Further investigation of the predicted compound–protein interactions was also done by finding supporting evidence from published literature, such as HSA being targeted by luteolin [41]. This result indicates that there are some compounds that might be considered as drugs in herbs. This also implies that the prediction model performs well and proteins that are not confirmed yet by any evidence can be candidates to have a relation with the corresponding efficacy group.

Table 4. Predicted compound–protein interactions by Jamu formulas. Compound ID is an identifier taken from PubChem CID (https://pubchem.ncbi.nlm.nih.gov, accessed on 20 August 2021) and KNApSAcK ID (http://kanaya.naist.jp/KNApSAcK_Family/, accessed on 20 August 2021). If the Compound ID cannot be found in PubChem or KNApSAcK databases, we assigned N/A.

No	Compound ID	Compound Name	Molecular Formula	UniProt ID	Targeted Protein	OMIM ID	Disease Description
				E1 Blood and Lymph Diseases			
1	N/A	(4Z)-1-(2,3,5-Trihydroxy-4-methylphenyl)dec-4-en-1-one	C17H24O4				
2	689043, C00000615	Caffeic acid	C9H8O4				
3	5317587,	Diacetoxy-[6]-gingerdiol	C21H32O6				
4	370, C00002647	Gallic acid	C7H6O5	P02768	Serum albumin	615999; 616000	Hyperthyroxinemia, familial dysalbuminemic; analbuminemia
5	5280445, C00000674	Luteolin	C15H10O6				
6	5280459, C00005373	Quercitrin	C21H20O11				
7	5281702, C00013329	Tricin	C17H14O7				
				Q92887	Canalicular multispecific organic anion transporter 1	237500	Dubin–Johnson syndrome
8	64945, C00003558	Ursolic acid	C30H48O3	Q9NPD5	Solute carrier organic anion transporter family member 1B3	237450	Hyperbilirubinemia, rotor type
				Q9Y6L6	Solute carrier organic anion transporter family member 1B1		Hyperbilirubinemia, rotor type
				P08185	Corticosteroid-binding globulin	611489	Corticosteroid-binding globulin deficiency
				P02768	Serum albumin	615999; 616000	Hyperthyroxinemia, familial dysalbuminemic; analbuminemia
9	73145, C00003738	beta-Amyrin	C30H50O	Q92887	Canalicular multispecific organic anion transporter 1	237500	Dubin–Johnson syndrome
10	222284, C00003672	beta-Sitosterol	C29H50O				Dubin–Johnson syndrome
				E3 The Digestive System			
1	519857, C00020146	1-epi-Cubenol	C15H26O				
2	N/A	Anisucumarin A	C20H20O8				
3	240, C00034452	Benzaldehyde	C7H6O				
4	6448, C00029844	Bornyl acetate	C12H20O2				
5	3314, C00000619	Eugenol	C10H12O2				
6	289151, C00003162	Longifolene	C15H24				
7	N/A	Morin-3-O-lyxoside	C20H18O11	P08183	Multidrug resistance protein 1	612244	Inflammatory bowel disease 13
8	985, C00001233	Palmitic acid	C16H32O2				
9	442402, C00003194	Thujopsene	C15H24				
10	12306047, C00029671	alpha-Muurolene	C15H24				
11	7460, C00003051	alpha-Phellandrene	C10H16				
12	111037, C00035043	alpha-Terpinyl acetate	C12H20O2				
13	12313020, C00020130	gamma-Muurolene	C15H24				
				E4 Female-Specific Diseases			
1	5280794, C00003674	Stigmasterol	C29H48O	P11511	Aromatase	139300; 613546	Aromatase excess syndrome; aromatase deficiency
				P03372	Estrogen receptor	615363	Estrogen resistance
				E7 Muscle and Bone			
1	10131321, C00055009	Coumaperine	C16H19NO2	P20309	Muscarinic acetylcholine receptor M3	100100	Prune belly syndrome
				E8 Nutritional and Metabolic Diseases			
				Q92887	Canalicular multispecific organic anion transporter 1	237500	Dubin–Johnson syndrome
1	3084331, C00020154	T-Muurolol	C15H26O	Q02318	Sterol 26-hydroxylase, mitochondrial	213700	Cerebrotendinous xanthomatosis
				P11473	Vitamin D3 receptor	277440	Rickets vitamin D-dependent 2A
				E10 Skin and Connective Tissue			
1	222284, C00003672	beta-Sitosterol	C29H50O	Q02318	Sterol 26-hydroxylase, mitochondrial	213700	Cerebrotendinous xanthomatosis
				E12 Mental and Behavioral Disorders			
1	6989, C00000155	Thymol	C10H14O	P08172	Muscarinic acetylcholine receptor M2	608516	Major depressive disorder
				Q13002	Glutamate receptor ionotropic, kainate 2	611092	Mental retardation, autosomal recessive 6

4. Conclusions and Future Works

We constructed classification–prediction models that predict the interactions between compounds and target proteins using a machine learning approach. The model was created by utilizing compound–protein interaction data obtained from open access databases, and the data were represented by a combination of fingerprint and amino acid sequences. The results showed very good prediction performances, around 90% when the compounds were transformed to MACCS fingerprint, amino acid sequences were transformed to AAC descriptor, and random forest was chosen as a classifier. In addition, some target proteins were predicted from potential compounds of Jamu formulas using the best model obtained in the previous step. By comparing the efficacy of predicted compounds and the relations of the targeted proteins with diseases, we found that some compounds might be considered as drug candidates. There are 27 compounds that can be validated by professional doctors, and those compounds belong to seven efficacy groups. This study is not only determines candidate drugs but also gives a better understanding of Jamu medicine at the omics level. Moreover, further validation of the results of this study can be performed by docking simulation between predicted compound–protein interactions or through in vivo and in vitro validation studies in the laboratory. We can also explore the supporting chemical or biological characteristics in predicted interactions, such as the similarity between the target compound and the known ligands of the predicted protein.

Author Contributions: Conceptualization, S.H.W. and M.A.-U.-A.; data curation, S.H.W.; formal analysis, S.H.W. and F.M.A.; funding acquisition, S.K.; investigation, M.H. and N.O.; methodology, S.H.W. and M.A.-U.-A.; supervision, I.B., S.K. and M.A.-U.-A.; writing—original draft, S.H.W.; writing—review and editing, M.A.-U.-A. All authors have read and agreed to the published version of the manuscript.

Funding: This work was supported by the World Class Professor Program Scheme A of the Ministry of Research, Technology and Higher Education of Indonesia and NAIST Big Data and Interdisciplinary Projects, Japan, and partially supported by the Ministry of Education, Culture, Sports, Science and Technology of Japan (16K07223 and 17K00406).

Data Availability Statement: Data used in this study were collected from previous studies and open access databases. Data are available from Computational Systems Biology Laboratory, NAIST, and Department of Computer Science of IPB University for researchers who meet the criteria (contact via correspondence authors).

Acknowledgments: We thank Husnawati and Nurida Dessalma Syahrania for validating predicted compound–protein interactions.

References

1. Harvey, A.L. Natural products in drug discovery. *Drug Discov. Today* **2008**, *13*, 894–901. [CrossRef] [PubMed]
2. Mu, C.; Sheng, Y.; Wang, Q.; Amin, A.; Li, X.; Xie, Y. Potential compound from herbal food of Rhizoma Polygonati for treatment of COVID-19 analyzed by network pharmacology: Viral and cancer signaling mechanisms. *J. Funct. Foods* **2021**, *77*, 10414. [CrossRef]
3. Chen, W.; Lim, C.E.D.; Kang, H.-J.; Liu, J. Chinese herbal medicines for the treatment of type A H1N1 influenza: A systematic review of randomized controlled trials. *PLoS ONE* **2011**, *6*, e028093. [CrossRef] [PubMed]
4. Safarzadeh, E.; Shotorbani, S.S.; Baradaran, B. Herbal medicine as inducers of apoptosis in cancer treatment. *Adv. Pharm. Bull.* **2014**, *4*, 421–427. [CrossRef]
5. Schippmann, U.; Leaman, D.J.; Cunningham, A.B. Impact of cultivation and gathering of medicinal plants on biodiversity: Global trends and issues. *Biodivers. Ecosyst. Approach Agric. For. Fish.* **2002**, 1–21. [CrossRef]
6. Schippmann, U.; Leaman, D.; Cunningham, A. A comparison of cultivation and wild collection of medicinal and aromatic plants under sustainability aspects. In *Medicinal and Aromatic Plants*; Springer: Dordrecht, The Netherlands, 2006; pp. 75–95. ISBN 9783540563914.
7. Hanafi, M.; Nina, A.; Fadia, Z.; Nurbaiti, N. *Indonesian Country Report on Traditional Medicine*; CSIR: New Delhi, India, 2006.
8. Furnham, A. Why do people choose and use complementary therapies. In *Complementary Medicine: An Objective Appraisal*; Ernst, E., Ed.; Butterworth-Heinemann: Oxford, UK, 1996; pp. 71–88.
9. Ernst, E. Herbal medicines put into context: Their use entails risks, but probably fewer than with synthetic drugs. *BMJ Br. Med.* **2003**, *327*, 881. [CrossRef]
10. Mahadevan, S.; Shah, S.L.; Marrie, T.J.; Slupsky, C.M. Analysis of metabolomic data using support vector machines. *Anal. Chem.* **2008**, *80*, 7562–7570. [CrossRef]
11. Wijaya, S.H.; Batubara, I.; Nishioka, T.; Altaf-Ul-Amin, M.; Kanaya, S. Metabolomic studies of Indonesian Jamu medicines: Prediction of Jamu efficacy and identification of important metabolites. *Mol. Inform.* **2017**, *36*, 1700050. [CrossRef] [PubMed]

2. Nasution, A.K.; Wijaya, S.H.; Kusuma, W.A. Prediction of drug-target interaction on Jamu formulas using machine learning approaches. In Proceedings of the ICACSIS 2019: 11th International Conference on Advanced Computer Science and Information Systems, Nusa Dua, Indonesia, 12–13 October 2019; pp. 169–174. [CrossRef]

3. Kanehisa, M.; Goto, S. KEGG: Kyoto encyclopedia of genes and genomes. *Nucleic Acids Res.* **2000**, *28*, 27–30. [CrossRef]

4. Law, V.; Knox, C.; Djoumbou, Y.; Jewison, T.; Guo, A.C.; Liu, Y.; MacIejewski, A.; Arndt, D.; Wilson, M.; Neveu, V.; et al. DrugBank 4.0: Shedding new light on drug metabolism. *Nucleic Acids Res.* **2014**, *42*, 1091–1097. [CrossRef] [PubMed]

5. Afendi, F.M.; Okada, T.; Yamazaki, M.; Hirai-Morita, A.; Nakamura, Y.; Nakamura, K.; Ikeda, S.; Takahashi, H.; Altaf-Ul-Amin, M.; Darusman, L.K.; et al. KNApSAcK family databases: Integrated metabolite-plant species databases for multifaceted plant research. *Plant Cell Physiol.* **2012**, *53*, e1. [CrossRef]

6. Bateman, A.; Martin, M.J.; O'Donovan, C.; Magrane, M.; Apweiler, R.; Alpi, E.; Antunes, R.; Arganiska, J.; Bely, B.; Bingley, M.; et al. UniProt: A hub for protein information. *Nucleic Acids Res.* **2015**, *43*, D204–D212. [CrossRef]

7. Hamosh, A.; Scott, A.F.; Amberger, J.S.; Bocchini, C.A.; McKusick, V.A. Online mendelian inheritance in man (OMIM), a knowledgebase of human genes and genetic disorders. *Nucleic Acids Res.* **2005**, *33*, 514–517. [CrossRef]

8. Yamanishi, Y.; Araki, M.; Gutteridge, A.; Honda, W.; Kanehisa, M. Prediction of drug-target interaction networks from the integration of chemical and genomic spaces. *Bioinformatics* **2008**, *24*, 232–240. [CrossRef]

9. Hattori, M.; Okuno, Y.; Goto, S.; Kanehisa, M. Development of a chemical structure comparison method for integrated analysis of chemical and genomic information in the metabolic pathways. *J. Am. Chem. Soc.* **2003**, *125*, 11853–11865. [CrossRef]

10. Smith, T.F.; Waterman, M.S. Identification of common molecular subsequences. *J. Mol. Biol.* **1981**, *147*, 195–197. [CrossRef]

11. Bleakley, K.; Yamanishi, Y. Supervised prediction of drug-target interactions using bipartite local models. *Bioinformatics* **2009**, *25*, 2397–2403. [CrossRef]

12. Gunn, S.R. *Support Vector Machines for Classification and Regression*; University of Southampton: Southampton, UK, 1998; Volume 14.

13. Bolton, E.E.; Wang, Y.; Thiessen, P.A.; Bryant, S.H. PubChem: Integrated platform of small molecules and biological activities. In *Annual Reports in Computational Chemistry*; Wheeler, R.A., Spellmeyer, D.C., Eds.; Elsevier: Amsterdam, The Netherlands, 2008; pp. 217–241.

14. Gunther, S.; Kuhn, M.; Dunkel, M.; Campillos, M.; Senger, C.; Petsalaki, E.; Ahmed, J.; Urdiales, E.G.; Gewiess, A.; Jensen, L.J.; et al. SuperTarget and matador: Resources for exploring drug-target relationships. *Nucleic Acids Res.* **2008**, *36*, 919–922. [CrossRef]

15. Wishart, D.S. DrugBank and its relevance to pharmacogenomics. *Pharmacogenomics* **2008**, *9*, 1155–1162. [CrossRef]

16. Durant, J.L.; Leland, B.A.; Henry, D.R.; Nourse, J.G. Reoptimization of MDL keys for use in drug discovery. *J. Chem. Inf. Comput. Sci.* **2002**, *42*, 1273–1280. [CrossRef]

17. Yap, C.W. PaDEL-descriptor: An open source software to calculate molecular descriptors and fingerprints. *J. Comput. Chem.* **2011**, *32*, 1466–1474. [CrossRef]

18. Xiao, N.; Cao, D.S.; Zhu, M.F.; Xu, Q.S. Protr/ProtrWeb: R package and web server for generating various numerical representation schemes of protein sequences. *Bioinformatics* **2015**, *31*, 1857–1859. [CrossRef]

19. Vapnik, V. *Statistical Learning Theory (Adaptive and Cognitive Dynamic Systems: Signal Processing, Learning, Communications and Control)*; John Wiley & Sons: Hoboken, NJ, USA, 1998; pp. 1–740.

20. Hussain, M.; Wajid, S.K.; Elzaart, A.; Berbar, M. A comparison of SVM kernel functions for breast cancer detection. In Proceedings of the 2011 8th International Conference on Computer Graphics, Imaging and Visualization (CGIV 2011), Singapore, 17–19 August 2011; pp. 145–150.

21. Breiman, L. Random forests. *Mach. Learn.* **2001**, *45*, 5–32. [CrossRef]

22. Díaz-Uriarte, R.; De Andres, S.A. Gene selection and classification of microarray data using random forest. *BMC Bioinform.* **2006**, *7*, 3. [CrossRef]

23. Jiang, R.; Tang, W.; Wu, X.; Fu, W. A random forest approach to the detection of epistatic interactions in case-control studies. *BMC Bioinform.* **2009**, *10*, S65. [CrossRef]

24. Duda, R.O.; Hart, P.E.; Stork, D.G. *Pattern Classification*; John Wiley & Sons: Hoboken, NJ, USA, 2012.

25. Deng, H.; Runger, G. Gene selection with guided regularized random forest. *Pattern Recognit.* **2013**, *46*, 3483–3489. [CrossRef]

26. Zhu, W.; Zeng, N.; Wang, N. Sensitivity, specificity, accuracy, associated confidence interval and ROC analysis with practical SAS® implementations. In Proceedings of the NESUG: Health Care and Life Sciences, Baltimore, MA, USA, 14–17 November 2010; pp. 1–9.

27. Wijaya, S.H.; Husnawati, H.; Afendi, F.M.; Batubara, I.; Darusman, L.K.; Altaf-Ul-Amin, M.; Sato, T.; Ono, N.; Sugiura, T.; Kanaya, S. Supervised clustering based on DPClusO: Prediction of plant-disease relations using Jamu formulas of KNApSAcK database. *Biomed Res. Int.* **2014**, *2014*, 831751. [CrossRef]

28. Dong, J.; Cao, D.S.; Miao, H.Y.; Liu, S.; Deng, B.C.; Yun, Y.H.; Wang, N.N.; Lu, A.P.; Zeng, W.B.; Chen, A.F. ChemDes: An integrated web-based platform for molecular descriptor and fingerprint computation. *J. Cheminform.* **2015**, *7*, 60. [CrossRef]

29. Meyer, D.; Dimitriadou, E.; Hornik, K.; Weingessel, A. *e1071: Misc Functions of the Department of Statistics (e1071)*; R Package Version 1(3); TU Wien: Vienna, Austria, 2014; pp. 1–62.

30. Yang, S.; Zhu, F.; Ling, X.; Liu, Q.; Zhao, P. Intelligent health care: Applications of deep learning in computational medicine. *Front. Genet.* **2021**, *12*, 607471. [CrossRef]

31. Jurasekova, Z.; Marconi, G.; Sanchez-Cortes, S.; Torreggiani, A. Spectroscopic and molecular modeling studies on the binding of the flavonoid luteolin and human serum albumin. *Biopolymers* **2009**, *91*, 917–927. [CrossRef]

Article

Shared Molecular Mechanisms of Hypertrophic Cardiomyopathy and Its Clinical Presentations: Automated Molecular Mechanisms Extraction Approach

Mila Glavaški [1,*] and Lazar Velicki [1,2]

[1] Faculty of Medicine, University of Novi Sad, Hajduk Veljkova 3, 21000 Novi Sad, Serbia; lazar.velicki@mf.uns.ac.rs

[2] Institute of Cardiovascular Diseases Vojvodina, Put Doktora Goldmana 4, 21204 Sremska Kamenica, Serbia

* Correspondence: milaglavaski@yahoo.com or milaglavaski@uns.ac.rs

Citation: Glavaški, M.; Velicki, L. Shared Molecular Mechanisms of Hypertrophic Cardiomyopathy and Its Clinical Presentations: Automated Molecular Mechanisms Extraction Approach. *Life* **2021**, *11*, 785. https://doi.org/10.3390/life11080785

Academic Editors: M. Altaf-Ul-Amin, Shigehiko Kanaya, Naoaki Ono and Ming Huang

Received: 16 June 2021
Accepted: 30 July 2021
Published: 3 August 2021

Publisher's Note: MDPI stays neutral with regard to jurisdictional claims in published maps and institutional affiliations.

Abstract: Hypertrophic cardiomyopathy (HCM) is the most common inherited cardiovascular disease with a prevalence of 1 in 500 people and varying clinical presentations. Although there is much research on HCM, underlying molecular mechanisms are poorly understood, and research on the molecular mechanisms of its specific clinical presentations is scarce. Our aim was to explore the molecular mechanisms shared by HCM and its clinical presentations through the automated extraction of molecular mechanisms. Molecular mechanisms were congregated by a query of the INDRA database, which aggregates knowledge from pathway databases and combines it with molecular mechanisms extracted from abstracts and open-access full articles by multiple machine-reading systems. The molecular mechanisms were extracted from 230,072 articles on HCM and 19 HCM clinical presentations, and their intersections were found. Shared molecular mechanisms of HCM and its clinical presentations were represented as networks; the most important elements in the intersections' networks were found, centrality scores for each element of each network calculated, networks with reduced level of noise generated, and cooperatively working elements detected in each intersection network. The identified shared molecular mechanisms represent possible mechanisms underlying different HCM clinical presentations. Applied methodology produced results consistent with the information in the scientific literature.

Keywords: hypertrophic cardiomyopathy; data mining; automated curation; molecular mechanisms; atrial fibrillation; sudden cardiac death; heart failure; left ventricular outflow tract obstruction; cardiac fibrosis; myocardial ischemia

1. Introduction

Hypertrophic cardiomyopathy (HCM) is the most common inherited cardiovascular disease [1–3], with an estimated prevalence of 1 in 500 people worldwide [1,3–5] and recent investigations suggesting an even greater prevalence [5,6]. It is characterized by increased left ventricular wall thickness that cannot be explained by abnormal loading conditions (e.g., hypertension) [1,2,7].

Mutations in genes that encode sarcomeric proteins are the primary molecular cause of HCM [3,8,9]. However, the genetic basis for HCM has proven to be more complex than originally postulated: 40–60% of HCM patients show mutations in one or more of the genes known to be associated with the disease, whereas for others, the cause remains unknown [10–12].

The clinical presentation of HCM varies widely [1,3,7,8]: some patients are asymptomatic [1,7,13], while others manifest symptomatic left ventricular outflow tract obstruction (LVOTO) [7,8], atrial fibrillation (AF) [3,8], sudden cardiac death (SCD) [3,7,13,14], or heart failure (HF) [1,3,10,11]. Pathophysiologic features of HCM include cardiomyocyte hypertrophy [15,16], cardiomyocyte disarray [16,17], myocardial remodeling [18,19], fibrosis [3,20,21], myocardial hypercontractility [22,23], impaired myocardial relaxation [20,24], myocardial stiffness [17,20], diastolic dysfunction [13,14,17], coronary microvascular dysfunction [25,26], and myocardial ischemia [25,27], but the underlying molecular mechanisms are poorly understood. Molecular determinants of the disease presentations are also still not known. Phenotypic expression of HCM may vary even within the same family [1] Despite active research, the consistent genotype-phenotype associations are still not known All these stress the importance of finding additional mechanisms and factors that direct the course and presentations of HCM, and propose all the molecular mechanisms standing between genetic basis and clinical presentations as crucial.

INDRA database [28] aggregates knowledge from pathway databases and combines it with information on molecular mechanisms extracted from abstracts and open-access full articles by multiple machine-reading systems. PubMed is one of the most important platforms for medical journal literature. To be indexed in PubMed, journals must meet certain review or selection criteria [29].

Our aim was to explore the shared molecular mechanisms of HCM and its clinical presentations through the automated extraction of molecular mechanisms.

2. Materials and Methods

2.1. Molecular Mechanisms Extraction

Molecular mechanisms were congregated using the INDRA database [28]. Molecular mechanisms from all PubMed articles published starting from 1 January 2010 were separately extracted in the form of INDRA statements [30] for HCM, cardiomyocyte hypertrophy, myofibrillar disarray, cardiomyocyte disarray, myocardial remodeling, cardiac remodeling, myocardial fibrosis, LVOTO, myocardial hypercontractility, impaired myocardial relaxation, impaired cardiac relaxation, myocardial stiffness, diastolic dysfunction, AF, SCD, coronary microvascular dysfunction, myocardial ischemia, HF, MACE, and rehospitalization. INDRA statements were found in the INDRA database by PubMed Identifier (PMIDs), using REST Client API. PMIDs were collected through the INDRA PubMed client [30] (which searches for articles on PubMed) using the following search terms: hypertrophic cardiomyopathy, cardiomyocyte hypertrophy, myofibrillar disarray, cardiomyocyte disarray, myocardial remodeling, cardiac remodeling, myocardial fibrosis, left ventricular outflow tract obstruction, myocardial hypercontractility, impaired myocardial relaxation, impaired cardiac relaxation, myocardial stiffness, diastolic dysfunction, atrial fibrillation sudden cardiac death, coronary microvascular dysfunction, myocardial ischemia, heart failure, major adverse cardiovascular events, and rehospitalization (use_text_word = True major_topic = True).

Subsequently, intersections of the sets consisting of INDRA statements for HCM and its clinical presentations were found.

2.2. Networks Generation

Each of the intersections (consisting of sets of INDRA statements) was transcribed to network table, imported to Cytoscape version 3.8.2 [31] for further analysis, and uploaded to NDEx v 2.5.0 [32–34].

2.3. Network Analysis

The most important nodes in intersections' networks were found using Cytoscape application Wk shell decomposition version 1.1.0 [35]. Rank and k-shell were calculated for every node of each network. The reliability of interactions was determined using Cytoscape PE-measure application version 1.0 [36]. Models with a reduced level of noise were generated and uploaded to NDEx. The nodes' centrality scores were determined using Cytoscape CytoHubba app version 0.1. [37]. Top elements for each centrality measure of each network were uploaded to NDEx. Cooperatively working elements (functional modules) were found using NCMine Cytoscape plugin version 1.3.0 [38]. All networks were analyzed as directed (with applied cliqueness threshold = 0.6, merge threshold = 0.6, dcliqueness threshold = 0.2, and cluster size threshold = 3).

3. Results

Molecular mechanisms in the form of 182,167 INDRA statements (representations of molecular mechanisms consisting molecular subject, object, and their interaction) were extracted from 230,072 articles on HCM and 19 HCM clinical presentations (Table 1).

Table 1. The number of articles on hypertrophic cardiomyopathy and its clinical presentations read automatically and the number of INDRA statements extracted.

Pathophysiologic Entity	Number of Articles Read Automatically	Number of INDRA Statements Extracted
hypertrophic cardiomyopathy	8111	7559
cardiomyocyte hypertrophy	1337	2500
myofibrillar disarray	51	356
cardiomyocyte disarray	11	22
myocardial remodeling	967	1500
cardiac remodeling	4572	5432
myocardial fibrosis	3634	4978
left ventricular outflow tract obstruction	1023	177
myocardial hypercontractility	3	3
impaired myocardial relaxation	31	33
impaired cardiac relaxation	12	28
myocardial stiffness	257	500
diastolic dysfunction	6342	6101
atrial fibrillation	54,117	25,842
sudden cardiac death	10,060	6770
coronary microvascular dysfunction	569	522
myocardial ischemia	19,637	19,078
heart failure	111,565	98,397
major adverse cardiovascular events	4700	1713
rehospitalization	3073	656

3.1. Network Analysis

3.1.1. Networks

Shared molecular mechanisms of HCM and its clinical presentations are represented as networks (Table 2). The networks differ notably in terms of the number of elements they contain.

The intersection of molecular interactions representing HCM and impaired cardiac relaxation contains only phosphorylation of SMAD Family Member 2 (SMAD2) and could not be displayed as a network. The intersection of HCM and myocardial hypercontractility contains no molecular interactions.

Table 2. Shared molecular mechanisms of hypertrophic cardiomyopathy and its clinical presentations

Pathophysiologic Entities	Link to the Network Representing Shared Molecular Mechanisms
hypertrophic cardiomyopathy, cardiomyocyte hypertrophy	https://bit.ly/39Yn90x (accessed on 1 August 2021)
hypertrophic cardiomyopathy, myofibrillar disarray	https://bit.ly/2PRnPOz (accessed on 1 August 2021)
hypertrophic cardiomyopathy, cardiomyocyte disarray	https://bit.ly/3wJsmmy (accessed on 1 August 2021)
hypertrophic cardiomyopathy, myocardial remodeling	https://bit.ly/2Q8dDkD (accessed on 1 August 2021)
hypertrophic cardiomyopathy, cardiac remodeling	https://bit.ly/31ZG3Qh (accessed on 1 August 2021)
hypertrophic cardiomyopathy, myocardial fibrosis	https://bit.ly/3fZX3hC (accessed on 1 August 2021)
hypertrophic cardiomyopathy, left ventricular outflow tract obstruction	https://bit.ly/3dN8G8R (accessed on 1 August 2021)
hypertrophic cardiomyopathy, impaired myocardial relaxation	https://bit.ly/322sU94 (accessed on 1 August 2021)
hypertrophic cardiomyopathy, myocardial stiffness	https://bit.ly/3mxmecq (accessed on 1 August 2021)
hypertrophic cardiomyopathy, diastolic dysfunction	https://bit.ly/3wHxRCn (accessed on 1 August 2021)
hypertrophic cardiomyopathy, atrial fibrillation	https://bit.ly/3d31kyT (accessed on 1 August 2021)
hypertrophic cardiomyopathy, sudden cardiac death	https://bit.ly/3wIN5ao (accessed on 1 August 2021)
hypertrophic cardiomyopathy, coronary microvascular dysfunction	https://bit.ly/31Xh2VN (accessed on 1 August 2021)
hypertrophic cardiomyopathy, myocardial ischemia	https://bit.ly/31YlC6a (accessed on 1 August 2021)
hypertrophic cardiomyopathy, heart failure	https://bit.ly/322UjI9 (accessed on 1 August 2021)
hypertrophic cardiomyopathy, major adverse cardiovascular events	https://bit.ly/3mvZZE1 (accessed on 1 August 2021)
hypertrophic cardiomyopathy, rehospitalization	https://bit.ly/3myyx8t (accessed on 1 August 2021)

3.1.2. The Most Important Nodes

The most important nodes for all networks are found (Supplementary Table S1). networks were presented as packed concentric rings sorted by the most important noc (Supplementary Figure S1).

3.1.3. Nodes' Centrality Scores

Centrality scores for each node of each network were calculated, and the top eleme for each centrality measure of each network were visualized (Table 3).

Table 3. Top nodes of each network ranked by centrality scores.

Network	Top Nodes Ranked by Centrality Scores
hypertrophic cardiomyopathy, cardiomyocyte hypertrophy	https://bit.ly/3fCs1Mq (accessed on 1 August 2021)
hypertrophic cardiomyopathy, myofibrillar disarray	https://bit.ly/2OgKHpM (accessed on 1 August 2021)
hypertrophic cardiomyopathy, cardiomyocyte disarray	https://bit.ly/31LLUsi (accessed on 1 August 2021)
hypertrophic cardiomyopathy, myocardial remodeling	https://bit.ly/3uj130t (accessed on 1 August 2021)
hypertrophic cardiomyopathy, cardiac remodeling	https://bit.ly/39CYWgj (accessed on 1 August 2021)
hypertrophic cardiomyopathy, myocardial fibrosis	https://bit.ly/3dc8HUA (accessed on 1 August 2021)
hypertrophic cardiomyopathy, left ventricular outflow tract obstruction	https://bit.ly/3cJRWzY (accessed on 1 August 2021)
hypertrophic cardiomyopathy, impaired myocardial relaxation	https://bit.ly/3dubAz7 (accessed on 1 August 2021)
hypertrophic cardiomyopathy, myocardial stiffness	https://bit.ly/2PpsZRM (accessed on 1 August 2021)
hypertrophic cardiomyopathy, diastolic dysfunction	https://bit.ly/2PQuwju (accessed on 1 August 2021)
hypertrophic cardiomyopathy, atrial fibrillation	https://bit.ly/2OhvNzE (accessed on 1 August 2021)
hypertrophic cardiomyopathy, sudden cardiac death	https://bit.ly/3ugy2CI (accessed on 1 August 2021)
hypertrophic cardiomyopathy, coronary microvascular dysfunction	https://bit.ly/3wiA5YR (accessed on 1 August 2021)
hypertrophic cardiomyopathy, myocardial ischemia	https://bit.ly/39Hexvk (accessed on 1 August 2021)
hypertrophic cardiomyopathy, heart failure	https://bit.ly/3uwQiYP (accessed on 1 August 2021)
hypertrophic cardiomyopathy, major adverse cardiovascular events	https://bit.ly/3fHsE7w (accessed on 1 August 2021)
hypertrophic cardiomyopathy, rehospitalization	https://bit.ly/3dzxLUy (accessed on 1 August 2021)

3.1.4. Reliability of Interactions

Networks with a reduced level of noise were generated (Table 4).

3.1.5. Cooperatively Working Elements

In each intersection network, cooperatively working elements (functional modules) were detected (Supplementary Table S2).

Table 4. Networks with different PE-values applied. PE-measure (the measure for interaction reliability) removes spurious interactions (below the value applied) and, thus, the level of noise in networks.

Network	Link to Networks with Different PE-Values Applied
hypertrophic cardiomyopathy, cardiomyocyte hypertrophy	https://bit.ly/3sMXLm1 (accessed on 1 August 2021)
hypertrophic cardiomyopathy, myofibrillar disarray	https://bit.ly/39Ebt2N (accessed on 1 August 2021)
hypertrophic cardiomyopathy, cardiomyocyte disarray	https://bit.ly/3cMpoGd (accessed on 1 August 2021)
hypertrophic cardiomyopathy, myocardial remodeling	https://bit.ly/3duSh8V (accessed on 1 August 2021)
hypertrophic cardiomyopathy, cardiac remodeling	https://bit.ly/3dBWPdU (accessed on 1 August 2021)
hypertrophic cardiomyopathy, myocardial fibrosis	https://bit.ly/324nVoj (accessed on 1 August 2021)
hypertrophic cardiomyopathy, left ventricular outflow tract obstruction	https://bit.ly/31GacUC (accessed on 1 August 2021)
hypertrophic cardiomyopathy, impaired myocardial relaxation	https://bit.ly/31MjKxv (accessed on 1 August 2021)
hypertrophic cardiomyopathy, myocardial stiffness	https://bit.ly/3sMZnfz (accessed on 1 August 2021)
hypertrophic cardiomyopathy, diastolic dysfunction	https://bit.ly/3cNFNu8 (accessed on 1 August 2021)
hypertrophic cardiomyopathy, atrial fibrillation	https://bit.ly/3mhXtRv (accessed on 1 August 2021)
hypertrophic cardiomyopathy, sudden cardiac death	https://bit.ly/3wxjGzZ (accessed on 1 August 2021)
hypertrophic cardiomyopathy, coronary microvascular dysfunction	https://bit.ly/3uhQQ4l (accessed on 1 August 2021)
hypertrophic cardiomyopathy, myocardial ischemia	https://bit.ly/31GmOLu (accessed on 1 August 2021)
hypertrophic cardiomyopathy, heart failure	https://bit.ly/3mi5i9U (accessed on 1 August 2021)
hypertrophic cardiomyopathy, major adverse cardiovascular events	https://bit.ly/2QZt66N (accessed on 1 August 2021)
hypertrophic cardiomyopathy, rehospitalization	https://bit.ly/2QYjedr (accessed on 1 August 2021)

3.2. Shared Molecular Elements and Pathways

3.2.1. Hypertrophic Cardiomyopathy and Structural Changes

The most important shared elements for cardiomyocyte hypertrophy and HCM we as follows: calcium; Ca^{2+}/calmodulin-dependent protein kinase II (CaMKII); *PLN* ge encoding phospholamban; protein kinase A (PKA), which is a master regulator of mo cAMP-dependent processes; protein kinase B (PKB), also known as AKT, which reg lates cellular survival and metabolism; AMP-activated protein kinase (AMPK), whi is involved in cellular energy homeostasis as a "cellular energy sensor"; and sirtuin encoded by *SIRT1* gene; nuclear factor of activated T-cells (NFAT), which is importa for immune response and is involved in the development of the cardiac system; *EDN* gene encoding endothelin 1 (ET-1), which is a potent vasoconstrictor; *AGT* gene, whi encodes angiotensinogen; collagen; multifunctional cytokine transforming growth fact β (TGF-β); signal transduction protein extracellular signal-regulated kinase (ERK); c population proliferation; and apoptosis.

The most important shared elements for myofibrillar disarray and HCM are act myosin complex, *MYL12A* gene, *MYBPC3* gene, ATP, mitogen-activated protein kina 7 (MAPK7) encoded by the *MAPK7* gene (MAP kinases are involved in many cellu processes), RAF proto-oncogene serine/threonine-protein kinase (RAF1)—a part of the ERK1/2 pathway as a MAP kinase encoded by *RAF1* gene, ERK, and *EDN1* gene encoding endothelin 1. The effects of immunosuppressant and calcineurin inhibitor cyclosporin A as well as MAP kinase cascade inhibitor PD98059 are also indicated.

In their pathophysiology, cardiomyocyte disarray and HCM share the mechanisms of contractile machinery components (actin, myosin complex, and enzyme ATPase); apoptosis-inhibiting mechanisms (B-cell lymphoma 2 gene, *BCL2*); the protein tyrosine phosphatase non-receptor type 11 (PTPN11), which inhibits the growth regulator—the mechanistic target of rapamycin, mTOR; and Src homology 2 domain-containing phosphatase 2 (Shp2), which is involved in cell growth and survival. The importance of the myosin heavy chain 7 gene, *MYH7*, is shown.

The common molecular elements of myocardial remodeling and HCM were as follows: calcium, CaMKII, *AGT* gene, which encodes angiotensinogen, angiotensin II, collagen, TGF-β, tumor necrosis factor (TNF), inflammatory response, cell population proliferation, and apoptosis.

Cardiac remodeling and HCM in their pathophysiology share calcium, AMPK (a "cellular energy sensor"), *AGT* gene encoding angiotensinogen, AKT (regulates cellular survival and metabolism), TGF-β (multifunctional cytokine), *SIRT1* gene encoding sirtuin 1 (SIRT1), collagen, actin, reactive oxygen species, cell population proliferation, and apoptosis.

By the most important nodes and ranked by centrality scores, the most important shared elements for myocardial fibrosis and HCM were calcium, TGF-β, collagen, *AGT* encoding angiotensinogen, angiotensin II, AMPK, cell population proliferation, inflammatory response, and apoptosis.

3.2.2. Hypertrophic Cardiomyopathy and Left Ventricular Outflow Tract Obstruction

LVOTO and HCM share calcium, TGF-β, *POSTN* gene encoding periostin (extracellular matrix protein with multiple functions), collagen, *PIMREG* gene and PIMREG protein (involved in metaphase-to-anaphase transition during mitosis), *SNCG* gene encoding gamma-synuclein (a member of the synuclein family of proteins, which were believed to be involved in the pathogenesis of neurodegenerative diseases and certain tumors), verapamil (calcium channel blocker), dobutamine (β1-agonist), mavacamten (MYK-461, inhibitor of cardiac myosin ATPase), systolic anterior motion, and death.

3.2.3. Hypertrophic Cardiomyopathy and Contractile Dysfunction

Impaired myocardial relaxation and HCM in their pathophysiology share nitric oxide (NO) and constitutive nitric oxide synthase (also known as nitic oxide synthase 3 (NOS3) or endothelial NOS) encoded by the *NOS3* gene as well as N, N-dimethylarginine, a direct endogenous inhibitor of NO synthases; interaction of phospholamban and ATP2A2 intracellular calcium pump; and collagen induction by TGF-β and PKA.

Molecules shared by myocardial stiffness and HCM were the *TTN* gene encoding titin, *RBM20* gene encoding RNA-binding protein that acts as a regulator of mRNA splicing of a subset of genes involved in cardiac development (regulates splicing of *TTN*), actin, TGF-β, *PRKCA* gene encoding protein kinase C-alpha (PKC-α), which was involved in diverse cellular signaling pathways, cyclic GMP-dependent protein kinase (PRKG), which involved in muscle relaxation, *RING1* gene encoding ring finger protein 1 (RING1), and *TRIM63* gene encoding tripartite motif-containing protein 63, which regulates the proteasomal degradation of muscle proteins.

Diastolic dysfunction and HCM share calcium, sodium, actin, troponin I, *TTN* gene encoding titin, *PLN* gene encoding phospholamban, cardiac myosin binding protein-C (cMyBP-C), CaMKII, TGF-β, PKA, AMPK, and apoptosis.

3.2.4. Hypertrophic Cardiomyopathy and Arrhythmia

Based on the most important nodes and centrality score ranks, the following elements were the most important shared elements for AF and HCM: sodium and calcium; CaMKII, which is important for calcium homeostasis in cardiomyocytes; *PLN* gene encoding phospholamban that inhibits the activity of ATPase sarcoplasmic/endoplasmic reticulum Ca^{2+} transporting 2 (*ATP2A2*—encodes one of the intracellular pumps that return calcium from the cytosol to the sarcoplasmic reticulum); *RYR2* gene encoding ryanodine receptor 2 (RyR2) (major mediator in calcium-induced calcium release from sarcoplasmic reticulum); *AGT* gene which encodes angiotensinogen, a precursor of angiotensin; junctophilin 2 (*JPH2*) gene which encodes a component of junctional complexes (it also plays a key role in calcium-induced calcium release); T-cell leukemia homeobox protein 2 (*TLX2*) gene; *PSMD4* gene which encodes component of the 26S proteasome, with its main role being the removal of misfolded or damaged proteins as well as proteins whose functions are no longer required. Both diseases share inflammatory response, apoptotic process, and death in their pathophysiology.

SCD and HCM share the following elements: sodium, calcium, *RYR2* gene encoding RyR2, CaMKII, actin, *MYL12A* gene, myosin complex, *TNNT1* gene encoding troponin T, troponin C, *TNNI3* gene encoding troponin I, ATP, PKA, *GJA1* gene encoding gap junction protein alpha 1 (connexin-43), *PLN* gene encoding phospholamban, *GSTK1* gene encoding glutathione S-transferase kappa 1, which belongs to a superfamily of enzymes for cellular detoxification, and death. The effect of the non-selective β adrenoceptor agonist isoprenaline is also indicated.

3.2.5. Hypertrophic Cardiomyopathy and Ischemia

The important common molecular mechanisms of coronary microvascular dysfunction and HCM are the serin/threonine-specific protein kinase Akt (also known as protein kinase B or PKB, this plays an important role in glucose metabolism, cell proliferation and apoptosis) and its activating phosphorylation site S473, structural sarcomeric protein titin, components of the phosphagen energy system (ATP decreased by creatine), glucose increased by insulin, calcium increased by sodium, and calcium increased by calcium. The effects of a few exogenous elements, such as antioxidant resveratrol (which activates the sirtuin 1 gene, *SIRT1*, regulator of whole-body lipid homeostasis), antimineralocorticoid spironolactone (inhibiting expression of mineralocorticoid receptor, encoded by nuclear receptor subfamily 3 group C member 2 gene, *NR3C2*), and insecticide pyraclofos (increasing calcium) are also indicated.

Myocardial ischemia and HCM share the following elements: calcium, ATP, AMP, PKA, AKT cross-talking with ERK, glucose, *INS*, *SIRT1* gene encoding sirtuin 1, NF-κB, TNF, reactive oxygen species, inflammatory response, and apoptosis.

3.2.6. Hypertrophic Cardiomyopathy and Endpoints

In their pathogenesis, HF and HCM share calcium, sodium, RYR2, components of contractile machinery and related genes actin, myosin complex, myosin light chain 12 (*MYL12A*) gene, troponin C, troponin T, *TNNI3* gene encoding troponin I, tropomyosin, myosin binding protein C3 (*MYBPC3*) gene, cMyBP-C, glucose, *INS* gene encoding insulin, reactive oxygen species, ATP, 3',5'-cyclic AMP (cAMP), ATPase, AMPK, sirtuin 1, TGF, collagen, *AGT* gene encoding angiotensinogen, *EDN1* gene encoding endothelin-1, ET, *GATA4* encoding transcription factor GATA-4, PKA, PKB (AKT), protein kinase C (PKC), p38 MAP kinase, beta adrenergic receptor genes (ADRBs), mechanistic target of rapamycin (mTOR), NF-κB, calmodulin, CaMKII, CaMKII-delta, *PLN* gene encoding phospholamban, *TTN* gene encoding titin, *CLEC3B* gene encoding tetranectin, *E2F1*, *PSMD4*, and *SMARC* genes, NFAT, β adrenoceptor agonist isoprenaline, inflammatory response, autophagy, cell population proliferation, apoptosis, and death.

Major adverse cardiovascular events (MACE) and HCM in their pathophysiology share calcium, (R)-lipoic acid (the most active isomer of a versatile antioxidant, alpha-lipoic acid), *NR3C2* gene encoding mineralocorticoid receptor, and ATPase (a class of enzymes that catalyze the hydrolysis of ATP to ADP). Apart from this, interesting elements for in the intersection and ranked as top nodes for some centrality measures are insulin, *PLN* gene encoding phospholamban, *PSMD4* gene, and TGF-β.

Rehospitalization and HCM share the following aspects: insulin receptor, glucose decrease mediated by insulin; ryanodine receptor; *PSMD4* gene related to increased death; *NLRP3* gene encoding regulator of immunity and inflammation cryopyrin (also known as angiotensin/vasopressin receptor AII/AVP-like), promoting proinflammatory cytokine interleukin 1 beta (IL-1B).

3.2.7. The Most Important Shared Elements and Pathways

The most important putative molecular elements and pathways are illustrated with corresponding HCM presentations (Figures 1 and 2).

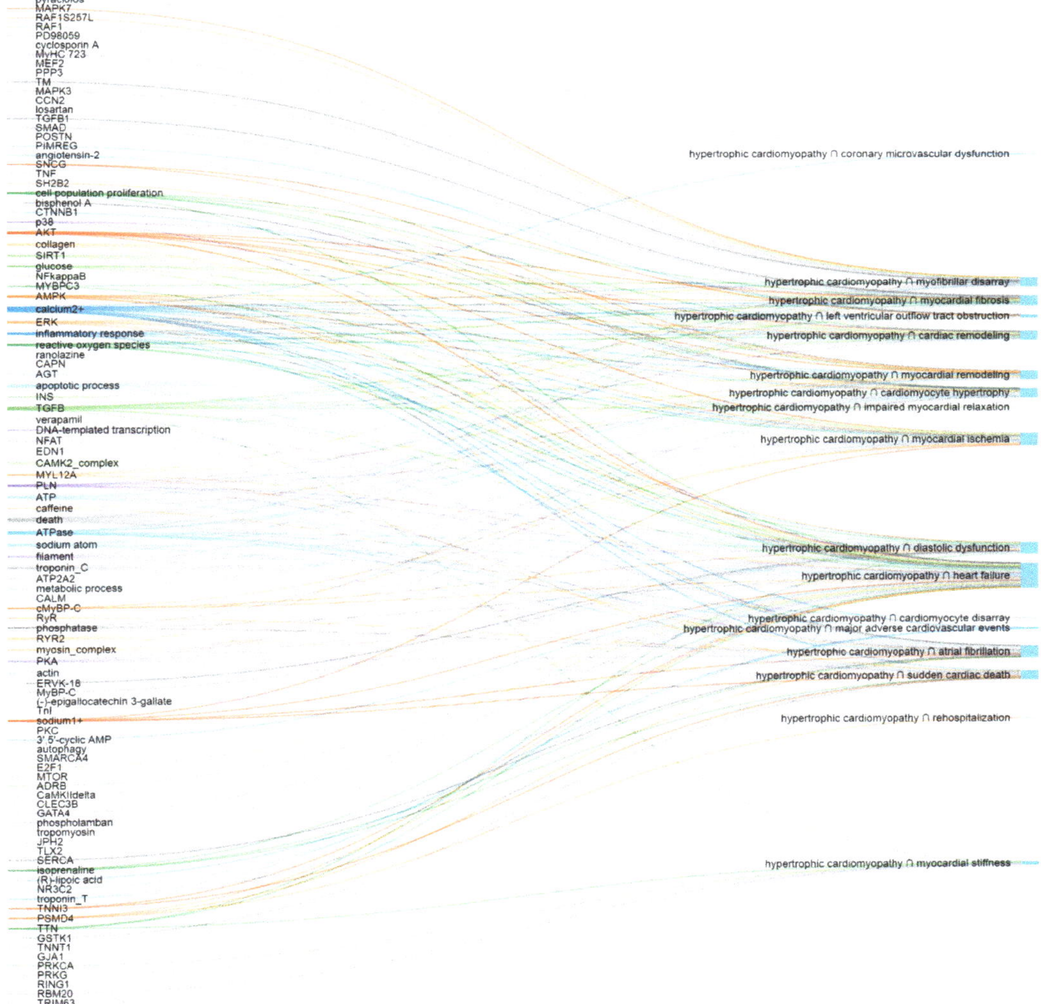

Figure 1. The most important putative molecular elements (**left**) and corresponding HCM presentations (**right**).

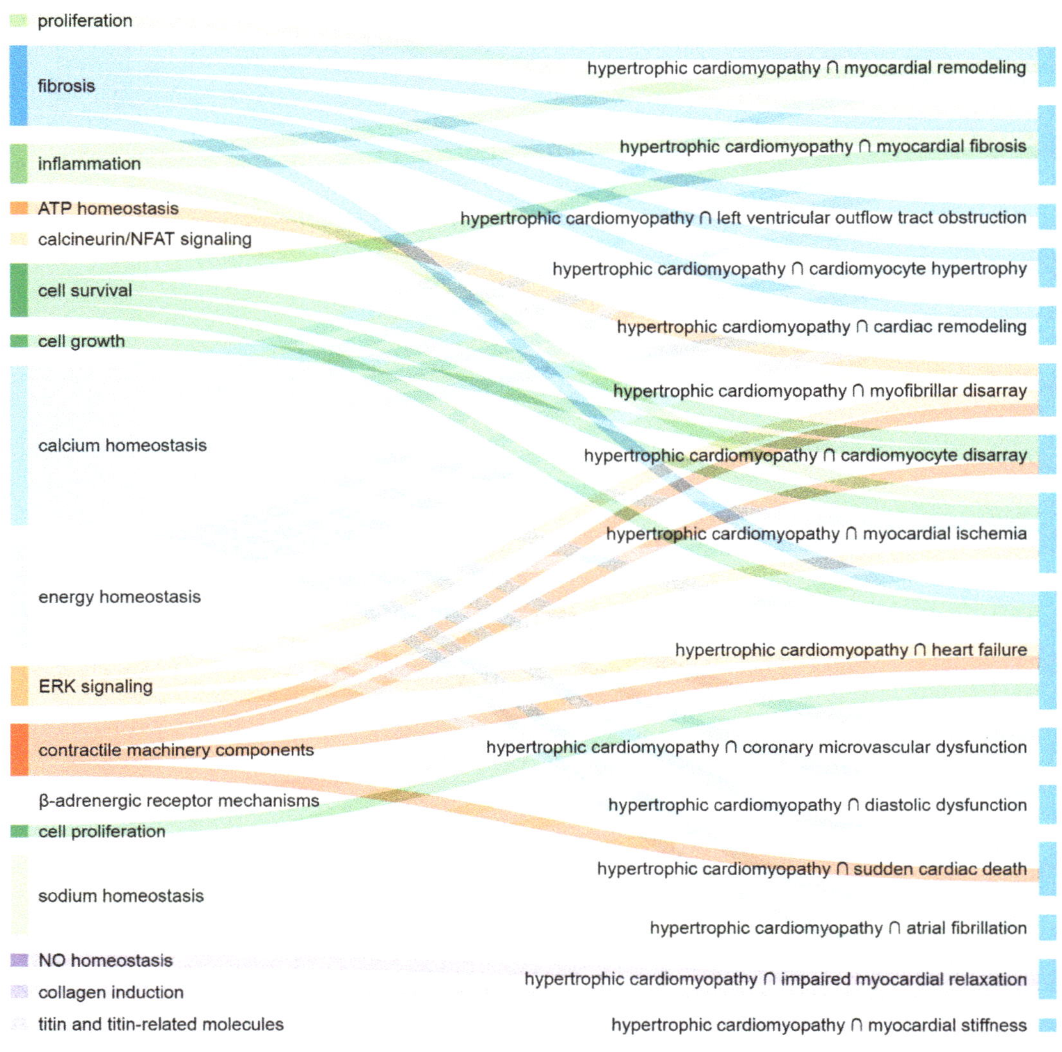

Figure 2. The most important putative pathways (**left**) and corresponding HCM presentations (**right**).

4. Discussion

To the best of our knowledge, this is the first study of shared molecular mechanisms of HCM and its clinical presentations.

Although there is much research about molecular mechanisms of HCM, research about molecular mechanisms in its specific clinical presentations is scarce. In the following literature review, we compared our results with evidence from preclinical and clinical literature.

4.1. Shared Molecular Elements and Pathways

4.1.1. Hypertrophic Cardiomyopathy and Structural Changes

The most important shared mechanisms of cardiomyocyte hypertrophy and HCM are those involved in fibrosis, calcium, and energy homeostasis. ET-1 strongly induces c diomyocyte hypertrophy in HCM-induced pluripotent stem-cells-derived cardiomyocytes, with ET-1 stimulation specifically inducing NFAT nuclear accumulation [39]. Angiotensin II induces cardiomyocyte hypertrophy [40–43].

The most important shared molecular mechanisms of myofibrillar disarray and HCM are contractile machinery components and those involved in ATP homeostasis, MAPK/ERK, and calcineurin/NFAT signaling. Frustaci et al. (2018) showed that in humans, mutation of sarcomeric α-actin is followed by fibrils disarray and hypertrophy with a disarray of cardiomyocytes, while dysfunction of cytoplasmic α-actin causes a disanchorage of myofibrils from the sarcolemmal membrane, followed by myofibrillolysis. The authors proposed that intercalated discs are particularly involved in this mutation, appearing irregular and fragmented, favoring cell disconnection [44]. Tanaka et al. (2014) showed that endothelin-1 induces myofibrillar disarray in HCM-induced pluripotent stem-cell-derived cardiomyocytes [39].

Our results indicate that the contractile machinery components and mechanisms involved in cell growth and survival are the most prominent mutual molecules and processes involved in cardiomyocyte disarray and HCM. Kraft et al. (2019) suggested that mutations in *MYH7* in heterozygous human HCM contribute to the development of cardiomyocyte disarray by burst-like heterogeneous expressions of both *MYH7* alleles (switched on and off in an independent and stochastic manner), which causes an imbalanced force generation going from cell to cell that disrupts the cardiac syncytium over time (stronger cells overstretch weaker cells) [45]. Schramm et al. (2012) showed that the *PTPN11* loss-of-function mutation Q510E-Shp2 causes cardiomyocyte disarray in HCM, with mTOR activation playing a critical role in the underlying mechanism [46]. James et al. (2000) demonstrated that one of the manifestations of cTnI^{146Gly} mutation in mice is cardiomyocyte disarray [47].

Our results indicate that calcium homeostasis, fibrosis, and inflammation mechanisms are the most important at the intersection of myocardial remodeling and HCM.

The most important shared elements of cardiac remodeling and HCM are implicated in fibrosis, calcium, and energy homeostasis. Freeman et al. (2001) showed that a high overexpression of β2-adrenergic receptor increases remodeling in HCM hearts and that inhibition of β-adrenergic receptor kinase (βARK) reverses hypertrophic remodeling in the HCM hearts [48]. Martins et al. (2015) suggested that the TNNC1-A8V mutant increases the calcium-binding affinity of the thin filament and elicits cellular remodeling [49]. Bi et al. (2021) showed that collagen cross-linking plays an important role in heart remodeling in human hypertrophic obstructive cardiomyopathy, which might be regulated mainly by lysyl oxidase (LOX) [50]. Roldán et al. (2008) suggested that the matrix metalloproteinases have an important role in cardiac remodeling in human HCM [51].

Shared molecules of myocardial fibrosis and HCM are those entangled in calcium and energy homeostasis, fibrosis, cell survival, proliferation, and inflammation. Ho et al. (2010) suggested that, in human HCM, sarcomere mutations trigger an early increase in collagen synthesis; this is initially balanced by degradation, but it exceeds degradation in overt HCM synthesis, resulting in myocardial fibrosis (i.e., collagen accumulation in HCM increases as the disease develops) [52]. Kawano et al. (2005) showed that valsartan (an angiotensin II type 1 receptor blocker) suppresses the synthesis of type I collagen in patients with HCM [53]. Arteaga et al. (2009) showed that myocardial fibrosis is prospectively associated with a worse prognosis in patients with HCM [54]. Further, Lim et al. (2001) showed that the blockade of angiotensin II (a known cardiotrophic factor) by losartan reverses myocardial fibrosis in a transgenic mouse model of human HCM [55].

4.1.2. Hypertrophic Cardiomyopathy and Left Ventricular Outflow Tract Obstruction

Scarce molecular mechanisms found at the intersection of LVOTO and HCM are indicative of an important role in calcium homeostasis and fibrosis. Bolca et al. (2002) showed that dobutamine induces dynamic LVOTO in patients with hypertrophic non-obstructive cardiomyopathy, proving that dobutamine stress echocardiography is a reliable tool for the diagnosis of dynamic left ventricular obstruction in patients with hypertrophic nonobstructive cardiomyopathy [56]. Mavacamten (a first-in-class cardiac myosin inhibitor has been evaluated as a promising new therapy in several clinical studies [57–59].

4.1.3. Hypertrophic Cardiomyopathy and Contractile Dysfunction

Although the molecular mechanisms found at the intersection of impaired myocardial relaxation and HCM are scarce, they indicate the leading role of NO homeostasis and a contribution of calcium homeostasis and collagen induction in their common pathogenesis. Cordts et al. (2019) suggested that higher N, N-dimethylarginine (also known as asymmetric dimethylarginine, ADMA) plasma concentrations might lead to a decreased NO production and an impaired myocardial relaxation in HCM patients [24].

Titin and titin-related molecules were found to be important in the intersection of myocardial stiffness and HCM. Higashikuse et al. (2019) suggested that titin mutations in HCM families can be incorporated into the sarcomere and impair TRIM63 (MURF1) binding, resulting in abnormal sarcomere stiffness [60].

Our results indicate that the contractile machinery components and mechanisms involved in calcium, sodium, and cellular energy homeostasis are the most prominent common molecules of diastolic dysfunction and HCM. Diastolic dysfunction in animal and human HCM is characterized by elevated myocardial activation at low diastolic calcium concentrations, i.e., high myofilament calcium-sensitivity [61–64]. In the majority of cases, the high basal (diastolic) myofilament activation is sufficient to slow the onset of ventricular relaxation and limit proper filling [62]. Sequeira et al. (2015) showed that tropomyosin's ability to block myosin-binding sites on actin is reduced in human HCM with thin-filament mutations, and the effect is exacerbated in human HCM samples by the low PKA phosphorylation of myofilament proteins. They also suggested that cMyBP-HCM-causing mutations reduce the accessibility of myosin for actin [65]. Teekakirikul et al. (2010) suggested that TGF-β signaling is implicated in progressive diastolic dysfunction in HCM [66]. Dweck et al. (2014) suggested that the inability to enhance myofilament relaxation through cardiac troponin I phosphorylation predisposes the heart to abnormal diastolic function [67]. Alves et al. (2015) proposed that troponin I may be an important target for the development of myofilament calcium desensitizers [68]. Further, Granzier et al. (2009) showed that the absence of PEVK region (one of the two major elastic elements of cardiac titin molecule) results in diastolic dysfunction [69].

4.1.4. Hypertrophic Cardiomyopathy and Arrhythmia

Our results suggest that the most important common mechanisms of AF and HCM are calcium and sodium homeostasis in cardiomyocytes. Bongini et al. (2016) suggested that RyR2 malfunction (probably by spontaneous sarcoplasmic reticulum calcium leakage) might represent a general pathophysiologic mechanism for AF initiation and maintenance in human HCM [70]. Nagai et al. (2007) found a significant association between the prevalence of AF and ACE polymorphism in patients with HCM [71].

Our results suggest that the most important shared molecular elements of SCD and HCM are contractile machinery components as well as sodium, calcium, and energy homeostasis mechanisms. Okuda et al. (2018) proved that CaMKII-mediated phosphorylation of RyR2 plays a crucial role in aberrant calcium release as a potent substrate of lethal arrhythmia in HCM-linked Troponin T-mutated hearts [72]. Alterations in calcium cling are triggers for cardiac arrhythmias—a serious clinical complication of HCM due to the potential to induce SCD [73]. On the other hand, calcium may be involved in the development of cardiac fibrosis, a potential substrate for cardiac arrhythmias and sudden death. In humans, mutations of calcium-related genes (RyR2 and calsequestrin 2) have been identified in families with a history of SCD [74]. Studies with HCM cardiomyocytes differentiated from patient-specific-induced pluripotent stem cells have confirmed that alterations of intracellular calcium handling are associated with arrhythmic events [Coppini et al. (2020) suggested that increased late sodium current (I_{NaL}) plays a central role in cellular arrhythmogenicity in HCM (which is confirmed by the antiarrhythmic effic of ranolazine) [76]. Parvatiyar et al. (2012) showed that *TNNC1* mutation A31S, which alters calcium handling, is associated with verified episodes of ventricular fibrillation and aborted SCD, probably due to altered calcium handling and electrophysiologic remodeling of the cardiomyocyte [77]. Additionally, Chung et al. (2011) found that frameshift mutation (c.363dupG) in Troponin C is associated with HCM and SCD [78]. Further, Fahed et al. (2020) showed that p.Arg21Cys mutation in *TNNI3* impairs calcium handling and results in a malignant HCM phenotype characterized by early-onset SCD [79]. HCM caused by mutations in the cardiac troponin T gene (*TNNT2*) has been associated with a high risk of SCD [80]. R58Q mutation of myosin regulatory light chain (*RLC*) is associated with SCD in HCM [81].

4.1.5. Hypertrophic Cardiomyopathy and Ischemia

We found only several common mechanisms in both coronary microvascular dysfunction and HCM. However, they showed the greatest importance of energy, calcium, and sodium homeostasis in the intersection of these two pathologies. We suggest that extracted interaction "glucose is increased/activated by insulin" refers to insulin-dependent glucose transport into cells, and "calcium increased/activated by calcium" is related to calcium-induced calcium release [82].

The most important elements of the intersection of molecular mechanisms of myocardial ischemia and HCM take part in calcium, sodium, and energy homeostasis, ERK signaling, inflammation, and cell survival.

4.1.6. Hypertrophic Cardiomyopathy and Endpoints

The most important shared molecular elements of HF and HCM are calcium, sodium, and energy homeostasis mechanisms, contractile machinery components, ERK signaling, β-adrenergic receptor mechanisms, and those entangled in fibrosis, cell proliferation, and survival. Mutations of *MYBPC3* gene are a major cause of human cardiomyopathy and associated HF [83]. *MYBPC3* mutations present a high risk for HF [84]. Kissopoulou et al. (2018) showed that homozygous missense *MYBPC3* Pro873His mutation in human HCM is associated with an increased risk of HF development [85]. Chronic administration of β-adrenergic agonists, such as isoproterenol, has been shown to aggravate HCM and induce HF in HCM models of disease [86].

We could not abstract the essence of the intersections of MACE and HCM or rehospitalization and HCM from the corresponding heterogeneous results, probably on account of the diverse pathologies underlying both MACE and rehospitalization.

4.1.7. Calcium in Hypertrophic Cardiomyopathy Presentations

Our results suggest that calcium is among the most important elements in almost all intersections of molecular pathways of HCM and its clinical presentations. Calcium is a key signaling molecule in the cardiac myocyte [74], and imbalances in calcium homeostasis have been described as key characteristics of HCM in numerous reports [73].

4.1.8. The Most Important Shared Elements and Pathways

As expected, at a high level, our results show that cardiomyocyte hypertrophy, myocardial and cardiac remodeling, and myocardial fibrosis; AF and SCD; coronary microvascular dysfunction and myocardial ischemia; myocardial ischemia and HF share similar molecular mechanisms, which is in line with clinical literature findings on HCM progression [87,88], arrhythmic nature and association between AF and SCD [89–91], ischemic nature and association between coronary microvascular dysfunction and myocardial ischemia [25,92,93], and association between myocardial ischemia and HF in HCM [94,95]. The results suggest a more isolated (distinctive) nature of myofibrillar and cardiomyocyte disarray, impaired myocardial relaxation, and myocardial stiffness, which might be, to some extent, a consequence of the relatively low number of articles available and statements extracted, which then reduce the ability to identify the most important molecular elements.

4.2. Non-Molecular Factors That Affect Clinical Presentations of HCM

Phenotypes of HCM are the consequences of complex interactions among a large number of determinants [96]. In addition to molecular mechanisms (including genetic factors), other factors can affect the clinical course and presentations of HCM. Environmental and lifestyle factors, most probably via epigenetic mechanisms, influence HCM phenotype [97–99]. These factors and their interaction in HCM have yet to be fully defined but might include microbial infection, diet [97], or exercise [96–98]. A study by Repetti et al. suggested that epigenetic and environmental factors, rather than background genetic variation, play a major role in hypertrophic remodeling [97].

Incomplete penetrance [96] and haploinsufficiency [84,99] also complicate interpretations of genotype-phenotype associations [96] and the prediction of clinical presentations. Phenotypic effects in cases of incomplete penetrance are even more responsive to the presence of other genetic and environmental factors. Cell-to-cell variability in gene expression and function also affect the HCM phenotype [96].

Physical factors like pressure changes, stretching, and changes in the generation of contraction force also influence the clinical course of HCM [96].

Other known and unknown factors might contribute to the development of different HCM clinical presentations as well.

By that means, this research, in its broad scope, is interesting for providing the potential of identification of molecular targets for environmental factors or lifestyle choices that could delay or change HCM progression.

4.3. General

All patients with HCM defined according to ESC guidelines [92] were included. No uniform exclusion criteria were applied.

Many molecular elements recognized as important in this research are non-specific and take part in different cardiac processes and diseases. Some of them might be compensatory mechanisms.

With the approach undertaken in the present study, we were able to detect shared mechanisms that might otherwise remain unnoticed. Although we cannot state that shared mechanisms determine or underlie the clinical presentation of HCM, these shared mechanisms have the potential to direct HCM processes or modify the nature of each disease state. Some of them might be novel therapeutic targets or contribute to the development of innovative strategies for treatment. This research also provides the potential to identify patients with specific or non-specific HCM molecular milieu patterns and with the preventability of certain complications or predisposition to side effects.

In silico studies of molecular interactions rarely provide final answers to questions. Nevertheless, very often, they produce a foundation for further research and initialize the generation of new questions and hypotheses. This work represents only the first step in the dissection of HCM pathogenesis, which could inspire and intensify future research. These results should be used after careful interpretation and critical evaluation of each element of interest in a particular use case.

Thus far, INDRA and the automated extraction of molecular mechanisms have been used in modeling p53 dynamics in response to DNA damage, adaptive drug resistance in BRAF-V600E-mutant melanomas, and the RAS signaling pathway [27].

Based on the literature review, the method applied has the potential to be beneficial in similar use cases. However, there is space for improvement of the technology and its implementation.

4.4. Limitations

The number of elements in networks (and sets) reflects the quantum of knowledge published on the topic rather than the complexity or granularity of the mechanisms themselves.

The automatic molecular mechanisms extraction approach is not specific (it extracts all molecular interactions from the article with the particular main topic), which is why each interaction should be considered critically.

The automatic extraction of molecular mechanisms sometimes extracts gene products with the name of the corresponding genes. Therefore, when evaluating an element with a gene name, it must be interpreted as the gene itself and/or its product.

Although automated extraction of molecular mechanisms creates a lot of clutter (e.g., elements not representing molecular mechanisms), we suppose that the nature of intersection removes most of the clutter (i.e., a piece of clutter should be present in two intersected sets to appear in results).

Both preclinical and clinical articles were included in the automatic molecular mechanisms extraction. Animal models do not fully replicate human HCM [100]. Our research lacks overall comorbidity information (it is source-article-specific).

4.5. Significance and Implications

This work collects and represents a quantum of knowledge about shared molecular mechanisms of HCM and its clinical presentations available today. Our results do not represent the final nor perfect dissection of HCM pathogenesis, yet they offer a transitional solution towards the next step in the research on HCM and its clinical presentations. It represents a wide foundation for further research, where new starting points could be found.

All pathways are presented in visual, and by that more intuitive, form, in one place. This work can be seen as a detailed review on the topic in the form of networks (instead of in the form of text) generated automatically (instead of by systematic literature inspection and writing). The pathways in the form of networks enable further analysis, for example, for in silico screening of new biomarkers and drug targets, as well as for predicting additional missing links and elements.

Shared pathways are commonly researched using different approaches [101–114]. The novelty in shared pathways research is the application of the new technology, automated molecular mechanisms extraction, to that task. In this research, we were also examining the reach of the technology used for automated extraction of molecular mechanisms from scientific medical literature. This approach is new in deciphering molecular mechanisms of HCM. Some parts of the methodology are taken over from the big data analysis field [14], and this research is one of the first attempts to analyze such massive data in the domain of this specific clinical entity.

This research also confirms that the results of usage of the technology are consistent with the information present in the scientific literature at a higher level, but also that there is a space for improvement of the technology and its implementation.

5. Conclusions

The most important molecular mechanisms that HCM shares with its clinical presentations are as follows: fibrosis, calcium and energy homeostasis (shared with cardiomyocyte hypertrophy and cardiac remodeling); contractile machinery components, ATP homeostasis, MAPK/ERK, and calcineurin/NFAT signaling (myofibrillar disarray); contractile machinery components and mechanisms involved in cell growth and survival (cardiomyocyte disarray); calcium homeostasis, fibrosis and inflammation mechanisms (myocardial remodeling); calcium and energy homeostasis, fibrosis, cell survival, proliferation and inflammation (myocardial fibrosis); calcium homeostasis and fibrosis (LVOTO); NO and calcium homeostasis, collagen induction (impaired myocardial relaxation); titin and titin-related molecules (myocardial stiffness); calcium and sodium homeostasis in cardiomyocytes (AF); contractile machinery components and mechanisms involved in calcium, sodium, and energy homeostasis (SCD and diastolic dysfunction); energy, calcium and sodium homeostasis mechanisms (coronary microvascular dysfunction); calcium, sodium and energy homeostasis; ERK signaling, inflammation and cell survival mechanisms (myocardial ischemia); calcium, sodium, and energy homeostasis mechanisms; contractile machinery components; ERK signaling; β-adrenergic receptor mechanisms; mechanisms entangled in fibrosis, cell proliferation and survival (HF). These mechanisms represen possible processes underlying different HCM clinical presentations, and some of them might be novel therapeutic targets.

This work collects and represents a quantum of knowledge about shared molecula mechanisms of HCM and its clinical presentations available today.

Applied methodology produced results consistent with the information in the scientifi literature at a higher level, but there is a space for improvement of the technology and it implementation.

Supplementary Materials: The following are available online at https://www.mdpi.com/article/1 .3390/life11080785/s1, Supplementary Table S1: The most important nodes in the networks, Table S: Cooperatively working elements (functional modules), Figure S1: The most important nodes i networks represented as packed concentric rings sorted by the most important nodes.

Author Contributions: Conceptualization, M.G. and L.V.; methodology, M.G. and L.V.; softwar M.G.; formal analysis, M.G.; investigation, M.G. and L.V.; resources, M.G. and L.V.; data curatio M.G.; writing—original draft preparation, M.G.; writing—review and editing, M.G. and L.V.; visua ization, M.G.; supervision, L.V.; project administration, M.G. and L.V.; funding acquisition, L.V. A authors have read and agreed to the published version of the manuscript.

Funding: This research has received funding from the European Union's Horizon 2020 Research ar Innovation Programme under grant agreement No. 777204 (www.silicofcm.eu, accessed on 1 Augu 2021). This article only reflects the author's view. The Commission is not responsible for any use th may be made of the information it contains.

Institutional Review Board Statement: Not applicable.

Informed Consent Statement: Not applicable.

Data Availability Statement: Data are contained within the article and supplementary material.

Acknowledgments: The authors are thankful to John Bachman, Harvard Medical School, for prov: ing access to the INDRA Database.

Conflicts of Interest: The authors declare no conflict of interest.

References

1. Sabater-Molina, M.; Pérez-Sánchez, I.; Hernández del Rincón, J.P.; Gimeno, J.R. Genetics of hypertrophic cardiomyopathy: review of current state. *Clin. Genet.* **2018**, *93*, 3–14. [CrossRef] [PubMed]
2. Firth, J. Cardiology: Hypertrophic cardiomyopathy. *Clin. Med.* **2019**, *19*, 61–63.
3. Geske, J.B.; Ommen, S.R.; Gersh, B.J. Hypertrophic cardiomyopathy: Clinical update. *JACC Heart Fail.* **2018**, *6*, 364–375. [CrossR
4. Deranek, A.E.; Klass, M.M.; Tardiff, J.C. Moving beyond simple answers to complex disorders in sarcomeric cardiomyopath: The role of integrated systems. *Pflügers Arch. Eur. J. Physiol.* **2019**, *471*, 661–671. [CrossRef] [PubMed]
5. Semsarian, C.; Ingles, J.; Maron, M.S.; Maron, B.J. New perspectives on the prevalence of hypertrophic cardiomyopathy. *J. / Coll. Cardiol.* **2015**, *65*, 1249–1254. [CrossRef] [PubMed]
6. Prondzynski, M.; Mearini, G.; Carrier, L. Gene therapy strategies in the treatment of hypertrophic cardiomyopathy. *Pflügers A Eur. J. Physiol.* **2019**, *471*, 807–815. [CrossRef] [PubMed]
7. Chiang, Y.P.; Shimada, Y.J.; Ginns, J.; Weiner, S.D.; Takayama, H. Septal myectomy for hypertrophic cardiomyopathy: Import surgical knowledge and technical tips in the era of increasing alcohol septal ablation. *Gen. Thorac. Cardiovasc. Surg.* **2018**, 192–200. [CrossRef]
8. Tuohy, C.V.; Kaul, S.; Song, H.K.; Nazer, B.; Heitner, S.B. Hypertrophic cardiomyopathy: The future of treatment. *Eur. J. Heart* **2020**, *22*, 228–240. [CrossRef]
9. Price, J.; Clarke, N.; Turer, A.; Quintana, E.; Mestres, C.; Huffman, L.; Peltz, M.; Wait, M.; Ring, W.S.; Jessen, M.; et al. Hypertrop obstructive cardiomyopathy: Review of surgical treatment. *Asian Cardiovasc. Thorac. Ann.* **2017**, *25*, 594–607. [CrossRef]
10. Cao, Y.; Zhang, P.Y. Review of recent advances in the management of hypertrophic cardiomyopathy. *Eur. Rev. Med. Pharmacol.* **2017**, *21*, 5207–5210.
11. Gómez, J.; Reguero, J.R.; Coto, E. The ups and downs of genetic diagnosis of hypertrophic cardiomyopathy. *Rev. Española Car* **2016**, *69*, 61–68. [CrossRef]

12. Tower-Rader, A.; Desai, M.Y. Phenotype–genotype correlation in hypertrophic cardiomyopathy. *Circ. Cardiovasc. Imaging* **2017**, *10*, e006066. [CrossRef]
13. Velicki, L.; Jakovljevic, D.G.; Preveden, A.; Golubovic, M.; Bjelobrk, M.; Ilic, A.; Stojsic, S.; Barlocco, F.; Tafelmeier, M.; Okwose, N.; et al. Genetic determinants of clinical phenotype in hypertrophic cardiomyopathy. *BMC Cardiovasc. Disord.* **2020**, *20*, 516. [CrossRef] [PubMed]
14. Smole, T.; Žunkovič, B.; Pičulin, M.; Kokalj, E.; Robnik-Šikonja, M.; Kukar, M.; Fotiadis, D.I.; Pezoulas, V.C.; Tachos, N.S.; Barlocco, F.; et al. A machine learning-based risk stratification model for ventricular tachycardia and heart failure in hypertrophic cardiomyopathy. *Comput. Biol. Med.* **2021**, *135*, 104648. [CrossRef]
15. Farrell, E.T.; Grimes, A.C.; de Lange, W.J.; Armstrong, A.E.; Ralphe, J.C. Increased postnatal cardiac hyperplasia precedes cardiomyocyte hypertrophy in a model of hypertrophic cardiomyopathy. *Front Physiol.* **2017**, *8*, 414. [CrossRef] [PubMed]
16. Ramachandra, C.J.A.; Mai Ja, K.P.M.; Lin, Y.H.; Shim, W.; Boisvert, W.A.; Hausenloy, D.J. Induced pluripotent stem cells for modelling energetic alterations in hypertrophic cardiomyopathy. *Cond. Med.* **2019**, *2*, 142–151.
17. MacIver, D.H.; Clark, A.L. Contractile dysfunction in sarcomeric hypertrophic cardiomyopathy. *J. Card. Fail.* **2016**, *22*, 731–737. [CrossRef]
18. Sukhacheva, T.V.; Chudinovskikh, Y.A.; Eremeeva, M.V.; Serov, R.A.; Bockeria, L.A. Proliferative potential of cardiomyocytes in hypertrophic cardiomyopathy: Correlation with myocardial remodeling. *Bull. Exp. Biol. Med.* **2016**, *162*, 160–169. [CrossRef] [PubMed]
19. Fernlund, E.; Gyllenhammar, T.; Jablonowski, R.; Carlsson, M.; Larsson, A.; Ärnlöv, J.; Liuba, P. Serum biomarkers of myocardial remodeling and coronary dysfunction in early stages of hypertrophic cardiomyopathy in the young. *Pediatr. Cardiol.* **2017**, *38*, 853–863. [CrossRef]
20. Ramachandra, C.J.A.; Kp, M.M.J.; Chua, J.; Hernandez-Resendiz, S.; Liehn, E.A.; Gan, L.M.; Michaëlsson, E.; Jonsson, M.K.B.; Ryden-Markinhuhta, K.; Bhat, R.V.; et al. Inhibiting cardiac myeloperoxidase alleviates the relaxation defect in hypertrophic cardiomyocytes. *Cardiovasc. Res.* **2021**, in press. [CrossRef]
21. Coppini, R.; Ferrantini, C.; Mugelli, A.; Poggesi, C.; Cerbai, E. Altered Ca^{2+} and Na^+ homeostasis in human hypertrophic cardiomyopathy: Implications for arrhythmogenesis. *Front. Physiol.* **2018**, *9*, 1391. [CrossRef]
22. Argirò, A.; Zampieri, M.; Berteotti, M.; Marchi, A.; Tassetti, L.; Zocchi, C.; Iannone, L.; Bacchi, B.; Cappelli, F.; Stefàno, P.; et al. Emerging Medical Treatment for Hypertrophic Cardiomyopathy. *J. Clin. Med.* **2021**, *10*, 951. [CrossRef]
23. Toepfer, C.N.; Wakimoto, H.; Garfinkel, A.C.; McDonough, B.; Liao, D.; Jiang, J.; Tai, A.C.; Gorham, J.M.; Lunde, I.G.; Lun, M.; et al. Hypertrophic cardiomyopathy mutations in MYBPC3 dysregulate myosin. *Sci. Transl. Med.* **2019**, *11*, eaat1199. [CrossRef] [PubMed]
24. Cordts, K.; Seelig, D.; Lund, N.; Carrier, L.; Böger, R.H.; Avanesov, M.; Tahir, E.; Schwedhelm, E.; Patten, M. Association of asymmetric dimethylarginine and diastolic dysfunction in patients with hypertrophic cardiomyopathy. *Biomolecules* **2019**, *9*, 277. [CrossRef] [PubMed]
25. Aguiar Rosa, S.; Rocha Lopes, L.; Fiarresga, A.; Ferreira, R.C.; Mota Carmo, M. Coronary microvascular dysfunction in hypertrophic cardiomyopathy: Pathophysiology, assessment, and clinical impact. *Microcirculation* **2021**, *28*, e12656. [CrossRef] [PubMed]
26. Yin, L.; Xu, H.Y.; Zheng, S.S.; Zhu, Y.; Xiao, J.X.; Zhou, W.; Yu, S.S.; Gong, L.G. 3.0 T magnetic resonance myocardial perfusion imaging for semi-quantitative evaluation of coronary microvascular dysfunction in hypertrophic cardiomyopathy. *Int. J. Cardiovasc. Imaging* **2017**, *33*, 1949–1959. [CrossRef] [PubMed]
27. Raphael, C.E.; Cooper, R.; Parker, K.H.; Collinson, J.; Vassiliou, V.; Pennell, D.J.; de Silva, R.; Hsu, L.Y.; Greve, A.M.; Nijjer, S.; et al. Mechanisms of myocardial ischemia in hypertrophic cardiomyopathy: Insights from wave intensity analysis and magnetic resonance. *J. Am. Coll. Cardiol.* **2016**, *68*, 1651–1660. [CrossRef]
28. INDRA Database. Available online: https://indra-db.readthedocs.io/en/latest/ (accessed on 31 May 2021).
29. Huh, S. How to add a journal to the international databases, Science Citation Index Expanded and MEDLINE. *Arch. Plast. Surg.* **2016**, *43*, 487. [CrossRef]
30. Gyori, B.M.; Bachman, J.A.; Subramanian, K.; Muhlich, J.L.; Galescu, L.; Sorger, P.K. From word models to executable models of signaling networks using automated assembly. *Mol. Syst. Biol.* **2017**, *13*, 954. [CrossRef]
31. Shannon, P.; Markiel, A.; Ozier, O.; Baliga, N.S.; Wang, J.T.; Ramage, D.; Amin, N.; Schwikowski, B.; Ideker, T. Cytoscape: A software environment for integrated models of biomolecular interaction networks. *Genome Res.* **2003**, *13*, 2498–2504. [CrossRef]
32. Pratt, D.; Chen, J.; Welker, D.; Rivas, R.; Pillich, R.; Rynkov, V.; Ono, K.; Miello, C.; Hicks, L.; Szalma, S.; et al. NDEx, the Network Data Exchange. *Cell Syst.* **2015**, *1*, 302–305. [CrossRef]
33. Pillich, R.T.; Chen, J.; Rynkov, V.; Welker, D.; Pratt, D. NDEx: A community resource for sharing and publishing of biological networks. *Methods Mol. Biol.* **2017**, *1558*, 271–301. [PubMed]
34. Pratt, D.; Chen, J.; Pillich, R.; Rynkov, V.; Gary, A.; Demchak, B.; Ideker, T. NDEx 2.0: A clearinghouse for research on cancer pathways. *Cancer Res.* **2017**, *77*, e58–e61. [CrossRef] [PubMed]
35. Cytoscape App Store, wk-shell-decomposition. Available online: http://apps.cytoscape.org/apps/wkshelldecomposition (accessed on 31 May 2021).
36. Zaki, N.; Efimov, D.; Berengueres, J. Protein complex detection using interaction reliability assessment and weighted clustering coefficient. *BMC Bioinform.* **2013**, *14*, 163. [CrossRef]

37. Chin, C.H.; Chen, S.H.; Wu, H.H.; Ho, C.W.; Ko, M.T.; Lin, C.Y. cytoHubba: Identifying hub objects and sub-networks from complex interactome. *BMC Syst. Biol.* **2014**, *8*, S11. [CrossRef]

38. Tadaka, S.; Kinoshita, K. NCMine: Core-peripheral based functional module detection using near-clique mining. *Bioinformatics* **2016**, *32*, 3454–3460. [CrossRef]

39. Tanaka, A.; Yuasa, S.; Mearini, G.; Egashira, T.; Seki, T.; Kodaira, M.; Kusumoto, D.; Kuroda, Y.; Okata, S.; Suzuki, T.; et al. Endothelin-1 induces myofibrillar disarray and contractile vector variability in hypertrophic cardiomyopathy-induced pluripotent stem cell-derived cardiomyocytes. *J. Am. Heart Assoc.* **2014**, *3*, e001263. [CrossRef] [PubMed]

40. Wu, T.; Wang, H.; Xin, X.; Yang, J.; Hou, Y.; Fang, M.; Lu, X.; Xu, Y. An MRTF-A–Sp1–PDE5 axis mediates angiotensin-II-induced cardiomyocyte hypertrophy. *Front. Cell Dev. Biol.* **2020**, *8*, 839. [CrossRef]

41. Yuan, Y.; Wang, J.; Chen, Q.; Wu, Q.; Deng, W.; Zhou, H.; Shen, D. Long non-coding RNA cytoskeleton regulator RNA (CYTOR) modulates pathological cardiac hypertrophy through miR-155-mediated IKKi signaling. *Biochim. Biophys. Acta Mol. Basis Dis* **2019**, *1865*, 1421–1427. [CrossRef]

42. Yu, X.J.; Huang, Y.Q.; Shan, Z.X.; Zhu, J.N.; Hu, Z.Q.; Huang, L.; Feng, Y.Q.; Geng, Q.S. MicroRNA-92b-3p suppresses angiotensin II-induced cardiomyocyte hypertrophy via targeting HAND2. *Life Sci.* **2019**, *232*, 116635. [CrossRef]

43. Shanmugam, P.; Valente, A.J.; Prabhu, S.D.; Venkatesan, B.; Yoshida, T.; Delafontaine, P.; Chandrasekar, B. Angiotensin-II type 1 receptor and NOX2 mediate TCF/LEF and CREB dependent WISP1 induction and cardiomyocyte hypertrophy. *J. Mol. Cell Cardiol.* **2011**, *50*, 928–938. [CrossRef] [PubMed]

44. Frustaci, A.; De Luca, A.; Guida, V.; Biagini, T.; Mazza, T.; Gaudio, C.; Letizia, C.; Russo, M.A.; Galea, N.; Chimenti, C. Novel α-actin gene mutation p.(Ala21Val) causing familial hypertrophic cardiomyopathy, myocardial noncompaction, and transmural crypts. Clinical-pathologic correlation. *J. Am. Heart. Assoc.* **2018**, *7*, e008068. [CrossRef]

45. Kraft, T.; Montag, J. Altered force generation and cell-to-cell contractile imbalance in hypertrophic cardiomyopathy. *Pflügers Arch. Eur. J. Physiol.* **2019**, *471*, 719–733. [CrossRef]

46. Schramm, C.; Fine, D.M.; Edwards, M.A.; Reeb, A.N.; Krenz, M. The *PTPN11* loss-of-function mutation Q510E-Shp2 causes hypertrophic cardiomyopathy by dysregulating mTOR signaling. *Am. J. Physiol. Heart Circ. Physiol.* **2012**, *302*, H231–H243. [CrossRef]

47. James, J.; Zhang, Y.; Osinska, H.; Sanbe, A.; Klevitsky, R.; Hewett, T.E.; Robbins, J. Transgenic modeling of a cardiac troponin mutation linked to familial hypertrophic cardiomyopathy. *Circ. Res.* **2000**, *87*, 805–811. [CrossRef] [PubMed]

48. Freeman, K.; Lerman, I.; Kranias, E.G.; Bohlmeyer, T.; Bristow, M.R.; Lefkowitz, R.J.; Iaccarino, G.; Koch, W.J.; Leinwand, L.A. Alterations in cardiac adrenergic signaling and calcium cycling differentially affect the progression of cardiomyopathy. *J. Clin. Investig.* **2001**, *107*, 967–974. [CrossRef]

49. Martins, A.S.; Parvatiyar, M.S.; Feng, H.Z.; Bos, J.M.; Gonzalez-Martinez, D.; Vukmirovic, M.; Turna, R.S.; Sanchez-Gonzalez, M.A.; Badger, C.D.; Zorio, D.A.R.; et al. In vivo analysis of troponin C knock-in (A8V) mice: Evidence that TNNC1 is a hypertrophic cardiomyopathy susceptibility gene. *Circ. Cardiovasc. Genet.* **2015**, *8*, 653–664. [CrossRef]

50. Bi, X.; Song, Y.; Song, Y.; Yuan, J.; Cui, J.; Zhao, S.; Qiao, S. Collagen cross-linking is associated with cardiac remodeling in hypertrophic obstructive cardiomyopathy. *J. Am. Heart Assoc.* **2021**, *10*, e017752. [CrossRef]

51. Roldán, V.; Marín, F.; Gimeno, J.R.; Ruiz-Espejo, F.; González, J.; Feliu, E.; García-Honrubia, A.; Saura, D.; de la Morena, G.; Valdés, M.; et al. Matrix metalloproteinases and tissue remodeling in hypertrophic cardiomyopathy. *Am. Heart J.* **2008**, *156*, 85–91. [CrossRef] [PubMed]

52. Ho, C.Y.; López, B.; Coelho-Filho, O.R.; Lakdawala, N.K.; Cirino, A.L.; Jarolim, P.; Kwong, R.; González, A.; Colan, S.D.; Seidman, J.G.; et al. Myocardial fibrosis as an early manifestation of hypertrophic cardiomyopathy. *N. Engl. J. Med.* **2010**, *363*, 552–563. [CrossRef]

53. Kawano, H.; Toda, G.; Nakamizo, R.; Koide, Y.; Seto, S.; Yano, K. Valsartan decreases type I collagen synthesis in patients with hypertrophic cardiomyopathy. *Circ. J.* **2005**, *69*, 1244–1248. [CrossRef]

54. Arteaga, E.; De Araújo, A.Q.; Bernstein, M.; Ramires, F.J.A.; Ianni, B.M.; Fernandes, F.; Mady, C. Prognostic value of the collagen volume fraction in hypertrophic cardiomyopathy. *Arq. Bras. Cardiol.* **2009**, *92*, 216–220.

55. Lim, D.S.; Lutucuta, S.; Bachireddy, P.; Youker, K.; Evans, A.; Entman, M.; Roberts, R.; Marian, A.J. Angiotensin II blockade reverses myocardial fibrosis in a transgenic mouse model of human hypertrophic cardiomyopathy. *Circulation* **2001**, *103*, 789–791. [CrossRef]

56. Bolca, O.; Özer, N.; Eren, M.; Dagdeviren, B.; Norgaz, T.; Akdemir, O.; Tezel, T. Dobutamine induced dynamic left ventricular outflow tract obstruction in patients with hypertrophic nonobstructive cardiomyopathy. *Tohoku J. Exp. Med.* **2002**, *198*, 79–88. [CrossRef]

57. Tower-Rader, A.; Ramchand, J.; Nissen, S.E.; Desai, M.Y. Mavacamten: A novel small molecule modulator of β-cardiac myosin for treatment of hypertrophic cardiomyopathy. *Expert Opin. Investig. Drugs* **2020**, *29*, 1171–1178. [CrossRef]

58. Heitner, S.B.; Jacoby, D.; Lester, S.J.; Owens, A.; Wang, A.; Zhang, D.; Lambing, J.; Lee, J.; Semigran, M.; Sehnert, A.J. Mavacamten treatment for obstructive hypertrophic cardiomyopathy: A clinical trial. *Ann. Intern. Med.* **2019**, *170*, 741–748. [CrossRef] [PubMed]

59. Olivotto, I.; Oreziak, A.; Barriales-Villa, R.; Abraham, T.P.; Masri, A.; Garcia-Pavia, P.; Saberi, S.; Lakdawala, N.K.; Wheeler, M.; Owens, A.; et al. Mavacamten for treatment of symptomatic obstructive hypertrophic cardiomyopathy (EXPLORER-HCM): A randomised, double-blind, placebo-controlled, phase 3 trial. *Lancet* **2020**, *396*, 759–769. [CrossRef]

0. Higashikuse, Y.; Mittal, N.; Arimura, T.; Yoon, S.H.; Oda, M.; Enomoto, H.; Kaneda, R.; Hattori, F.; Suzuki, T.; Kawakami, A.; et al. Perturbation of the titin/MURF1 signaling complex is associated with hypertrophic cardiomyopathy in a fish model and in human patients. *Dis. Model Mech.* **2019**, *12*, dmm041103. [CrossRef]

1. Abraham, T.P.; Jones, M.; Kazmierczak, K.; Liang, H.Y.; Pinheiro, A.C.; Wagg, C.S.; Lopaschuk, G.D.; Szczesna-Cordary, D. Diastolic dysfunction in familial hypertrophic cardiomyopathy transgenic model mice. *Cardiovasc. Res.* **2009**, *82*, 84–92. [CrossRef] [PubMed]

2. Sequeira, V.; Bertero, E.; Maack, C. Energetic drain driving hypertrophic cardiomyopathy. *FEBS Lett.* **2019**, *593*, 1616–1626. [CrossRef]

3. Wijnker, P.J.M.; Sequeira, V.; Kuster, D.W.D.; Velden, J.V. Hypertrophic cardiomyopathy: A vicious cycle triggered by sarcomere mutations and secondary disease hits. *Antioxid. Redox Signal.* **2019**, *31*, 318–358. [CrossRef]

4. Wu, H.; Yang, H.; Rhee, J.W.; Zhang, J.Z.; Lam, C.K.; Sallam, K.; Chang, A.C.Y.; Ma, N.; Lee, J.; Zhang, H.; et al. Modelling diastolic dysfunction in induced pluripotent stem cell-derived cardiomyocytes from hypertrophic cardiomyopathy patients. *Eur. Heart J.* **2019**, *40*, 3685–3695. [CrossRef]

5. Sequeira, V.; Najafi, A.; Wijnker, P.J.M.; Dos Remedios, C.G.; Michels, M.; Kuster, D.W.D.; van der Velden, J. ADP-stimulated contraction: A predictor of thin-filament activation in cardiac disease. *Proc. Natl. Acad. Sci. USA* **2015**, *112*, E7003–E7012. [CrossRef] [PubMed]

6. Teekakirikul, P.; Eminaga, S.; Toka, O.; Alcalai, R.; Wang, L.; Wakimoto, H.; Nayor, M.; Konno, T.; Gorham, J.M.; Wolf, C.M.; et al. Cardiac fibrosis in mice with hypertrophic cardiomyopathy is mediated by non-myocyte proliferation and requires Tgf-β. *J. Clin. Investig.* **2010**, *120*, 3520–3529. [CrossRef] [PubMed]

7. Dweck, D.; Sanchez-Gonzalez, M.A.; Chang, A.N.; Dulce, R.A.; Badger, C.D.; Koutnik, A.P.; Ruiz, E.L.; Griffin, B.; Liang, J.; Kabbaj, M.; et al. Long term ablation of protein kinase A (PKA)-mediated cardiac troponin I phosphorylation leads to excitation-contraction uncoupling and diastolic dysfunction in a knock-in mouse model of hypertrophic cardiomyopathy. *J. Biol. Chem.* **2014**, *289*, 23097–23111. [CrossRef]

8. Alves, M.L.; Dias, F.A.L.; Gaffin, R.D.; Simon, J.N.; Montminy, E.M.; Biesiadecki, B.J.; Hinken, A.C.; Warren, C.M.; Utter, M.S.; Davis, R.T.; et al. Desensitization of myofilaments to Ca^{2+} as a therapeutic target for hypertrophic cardiomyopathy with mutations in thin filament proteins. *Circ. Cardiovasc. Genet.* **2014**, *7*, 132–143. [CrossRef]

9. Granzier, H.L.; Radke, M.H.; Peng, J.; Westermann, D.; Nelson, O.L.; Rost, K.; King, N.M.P.; Yu, Q.; Tschöpe, C.; McNabb, M.; et al. Truncation of titin's elastic PEVK region leads to cardiomyopathy with diastolic dysfunction. *Circ. Res.* **2009**, *105*, 557–564. [CrossRef] [PubMed]

Bongini, C.; Ferrantini, C.; Girolami, F.; Coppini, R.; Arretini, A.; Targetti, M.; Bardi, S.; Castelli, G.; Torricelli, F.; Cecchi, F.; et al. Impact of genotype on the occurrence of atrial fibrillation in patients with hypertrophic cardiomyopathy. *Am. J. Cardiol.* **2016**, *117*, 1151–1159. [CrossRef] [PubMed]

Nagai, T.; Ogimoto, A.; Okayama, H.; Ohtsuka, T.; Shigematsu, Y.; Hamada, M.; Miki, T.; Higaki, J. A985G polymorphism of the endothelin-2 gene and atrial fibrillation in patients with hypertrophic cardiomyopathy. *Circ. J.* **2007**, *71*, 1932–1936. [CrossRef]

Okuda, S.; Sufu-Shimizu, Y.; Kato, T.; Fukuda, M.; Nishimura, S.; Oda, T.; Kobayashi, S.; Yamamoto, T.; Morimoto, S.; Yano, M. CaMKII-mediated phosphorylation of RyR2 plays a crucial role in aberrant Ca^{2+} release as an arrhythmogenic substrate in cardiac troponin T-related familial hypertrophic cardiomyopathy. *Biochem. Biophys. Res. Commun.* **2018**, *496*, 1250–1256. [CrossRef]

Lan, F.; Lee, A.S.; Liang, P.; Sanchez-Freire, V.; Nguyen, P.K.; Wang, L.; Han, L.; Yen, M.; Wang, Y.; Sun, N.; et al. Abnormal calcium handling properties underlie familial hypertrophic cardiomyopathy pathology in patient-specific induced pluripotent stem cells. *Cell Stem Cell* **2013**, *12*, 101–113. [CrossRef] [PubMed]

Tsoutsman, T.; Lam, L.; Semsarian, C. Genes, calcium and modifying factors in hypertrophic cardiomyopathy. *Clin. Exp. Pharmacol. Physiol.* **2006**, *33*, 139–145. [CrossRef]

Han, L.; Li, Y.; Tchao, J.; Kaplan, A.D.; Lin, B.; Li, Y.; Mich-Basso, J.; Lis, A.; Hassan, N.; London, B.; et al. Study familial hypertrophic cardiomyopathy using patient-specific induced pluripotent stem cells. *Cardiovasc. Res.* **2014**, *104*, 258–269. [CrossRef]

Coppini, R.; Santini, L.; Olivotto, I.; Ackerman, M.J.; Cerbai, E. Abnormalities in sodium current and calcium homoeostasis as drivers of arrhythmogenesis in hypertrophic cardiomyopathy. *Cardiovasc. Res.* **2020**, *116*, 1585–1599. [CrossRef]

Parvatiyar, M.S.; Landstrom, A.P.; Figueiredo-Freitas, C.; Potter, J.D.; Ackerman, M.J.; Pinto, J.R. A mutation in TNNC1-encoded cardiac troponin C, TNNC1-A31S, predisposes to hypertrophic cardiomyopathy and ventricular fibrillation. *J. Biol. Chem.* **2012**, *287*, 31845–31855. [CrossRef]

Chung, W.K.; Kitner, C.; Maron, B.J. Novel frameshift mutation in Troponin C (TNNC1) associated with hypertrophic cardiomyopathy and sudden death. *Cardiol. Young* **2011**, *21*, 345–348. [CrossRef] [PubMed]

Fahed, A.C.; Nemer, G.; Bitar, F.F.; Arnaout, S.; Abchee, A.B.; Batrawi, M.; Khalil, A.; Abou Hassan, O.K.; DePalma, S.R.; McDonough, B.; et al. Founder mutation in N terminus of cardiac troponin I causes malignant hypertrophic cardiomyopathy. *Circ. Genom. Precis. Med.* **2020**, *13*, 444–452. [CrossRef] [PubMed]

Pasquale, F.; Syrris, P.; Kaski, J.P.; Mogensen, J.; McKenna, W.J.; Elliott, P. Long-term outcomes in hypertrophic cardiomyopathy caused by mutations in the cardiac troponin T gene. *Circ. Cardiovasc. Genet.* **2012**, *5*, 10–17. [CrossRef]

Karabina, A.; Kazmierczak, K.; Szczesna-Cordary, D.; Moore, J.R. Myosin regulatory light chain phosphorylation enhances cardiac β-myosin in vitro motility under load. *Arch. Biochem. Biophys.* **2015**, *580*, 14–21. [CrossRef]

Roderick, H.L.; Berridge, M.J.; Bootman, M.D. Calcium-induced calcium release. *Curr. Biol.* **2003**, *13*, R425. [CrossRef]

83. Knöll, R. Myosin binding protein C: Implications for signal-transduction. *J. Muscle Res. Cell Motil.* **2012**, *33*, 31–42. [CrossRef] [PubMed]
84. Arif, M.; Nabavizadeh, P.; Song, T.; Desai, D.; Singh, R.; Bazrafshan, S.; Kumar, M.; Wang, Y.; Gilbert, R.J.; Dhandapany, P.S.; et al. Genetic, clinical, molecular, and pathogenic aspects of the South Asian–specific polymorphic *MYBPC3^{Δ25bp}* variant. *Biophys. Rev.* **2020**, *12*, 1065–1084. [CrossRef] [PubMed]
85. Kissopoulou, A.; Trinks, C.; Green, A.; Karlsson, J.E.; Jonasson, J.; Gunnarsson, C. Homozygous missense MYBPC3 Pro873His mutation associated with increased risk for heart failure development in hypertrophic cardiomyopathy. *ESC Hear Fail.* **2018**, *5*, 716–723. [CrossRef]
86. Li, X.; Lu, W.J.; Li, Y.; Wu, F.; Bai, R.; Ma, S.; Dong, T.; Zhang, H.; Lee, A.S.; Wang, Y.; et al. MLP-deficient human pluripotent stem cell derived cardiomyocytes develop hypertrophic cardiomyopathy and heart failure phenotypes due to abnormal calcium handling. *Cell Death Dis.* **2019**, *10*, 610. [CrossRef] [PubMed]
87. Schirone, L.; Forte, M.; Palmerio, S.; Yee, D.; Nocella, C.; Angelini, F.; Pagano, F.; Schiavon, S.; Bordin, A.; Carrizzo, A.; et al. A review of the molecular mechanisms underlying the development and progression of cardiac remodeling. *Oxid. Med. Cell. Longev.* **2017**, *2017*, 3920195. [CrossRef]
88. Liu, T.; Song, D.; Dong, J.; Zhu, P.; Liu, J.; Liu, W.; Ma, X.; Zhao, L.; Ling, S. Current Understanding of the pathophysiology of myocardial fibrosis and its quantitative assessment in heart failure. *Front Physiol.* **2017**, *8*, 238. [CrossRef]
89. Jordà, P.; García-Álvarez, A. Hypertrophic cardiomyopathy: Sudden cardiac death risk stratification in adults. *Glob. Cardiol. Sci. Pract.* **2018**, *2018*, 25. [CrossRef]
90. Waldmann, V.; Jouven, X.; Narayanan, K.; Piot, O.; Chugh, S.S.; Albert, C.M.; Marijon, E. Association between atrial fibrillation and sudden cardiac death. *Circ. Res.* **2020**, *127*, 301–309. [CrossRef]
91. O'Mahony, C.; Elliott, P.; McKenna, W. Sudden cardiac death in hypertrophic cardiomyopathy. *Circ. Arrhythm. Electrophysiol.* **2013**, *6*, 443–451. [CrossRef]
92. Petersen, S.E.; Jerosch-Herold, M.; Hudsmith, L.E.; Robson, M.D.; Francis, J.M.; Doll, H.A.; Selvanayagam, J.B.; Neubauer, S.; Watkins, H. Evidence for microvascular dysfunction in hypertrophic cardiomyopathy. *Circulation* **2007**, *115*, 2418–2425. [CrossRef]
93. Cecchi, F.; Olivotto, I.; Gistri, R.; Lorenzoni, R.; Chiriatti, G.; Camici, P.G. Coronary microvascular dysfunction and prognosis hypertrophic cardiomyopathy. *N. Engl. J. Med.* **2003**, *349*, 1027–1035. [CrossRef]
94. Maron, M.S.; Olivotto, I.; Maron, B.J.; Prasad, S.K.; Cecchi, F.; Udelson, J.E.; Camici, P.G. The case for myocardial ischemia hypertrophic cardiomyopathy. *J. Am. Coll. Cardiol.* **2009**, *54*, 866–875. [CrossRef]
95. Raphael, C.E.; Mitchell, F.; Kanaganayagam, G.S.; Liew, A.C.; Di Pietro, E.; Vieira, M.S.; Kanapeckaite, L.; Newsome, S.; Gregson, J.; Owen, R.; et al. Cardiovascular magnetic resonance predictors of heart failure in hypertrophic cardiomyopathy: The role myocardial replacement fibrosis and the microcirculation. *J. Cardiovasc. Magn. Reson.* **2021**, *23*, 26. [CrossRef]
96. Marian, A.J.; Braunwald, E. Hypertrophic cardiomyopathy: Genetics, pathogenesis, clinical manifestations, diagnosis, and therapy. *Circ. Res.* **2017**, *121*, 749–770. [CrossRef]
97. Repetti, G.G.; Kim, Y.; Pereira, A.C.; Ingles, J.; Russell, M.W.; Lakdawala, N.K.; Ho, C.Y.; Day, S.; Semsarian, C.; McDonough, B.; et al. Discordant clinical features of identical hypertrophic cardiomyopathy twins. *Proc. Natl. Acad. Sci. USA* **2021**, *118*, e2021717118. [CrossRef]
98. Pérez-Sánchez, I.; Romero-Puche, A.J.; García-Molina Sáez, E.; Sabater-Molina, M.; López-Ayala, J.M.; Muñoz-Esparza, C.; López-Cuenca, D.; de la Morena, G.; Castro-García, F.J.; Gimeno-Blanes, J.R. Factors influencing the phenotypic expression hypertrophic cardiomyopathy in genetic carriers. *Rev. Esp. Cardiol.* **2018**, *71*, 146–154. [CrossRef] [PubMed]
99. Barefield, D.; Kumar, M.; Gorham, J.; Seidman, J.G.; Seidman, C.E.; de Tombe, P.P.; Sadayappan, S. Haploinsufficiency of *MYBPC3* exacerbates the development of hypertrophic cardiomyopathy in heterozygous mice. *J. Mol. Cell Cardiol.* **2015**, *79*, 234–243. [CrossRef] [PubMed]
100. Ueda, Y.; Stern, J.A. A one health approach to hypertrophic cardiomyopathy. *Yale J. Biol. Med.* **2017**, *90*, 433–448. [PubMed]
101. Tye, C.; Runicles, A.K.; Whitehouse, A.J.O.; Alvares, G.A. Characterizing the interplay between autism spectrum disorder and comorbid medical conditions: An integrative review. *Front. Psychiatry* **2018**, *9*, 751. [CrossRef]
102. Hoyt, C.T.; Domingo-Fernández, D.; Balzer, N.; Güldenpfennig, A.; Hofmann-Apitius, M. A systematic approach for identifying shared mechanisms in epilepsy and its comorbidities. *Database* **2018**, *2018*, bay050. [CrossRef]
103. Ko, Y.; Cho, M.; Lee, J.S.; Kim, J. Identification of disease comorbidity through hidden molecular mechanisms. *Sci. Rep.* **2016**, 39433. [CrossRef]
104. Meng, Z.Q.; Wu, J.R.; Zhu, Y.L.; Zhou, W.; Fu, C.G.; Liu, X.K.; Liu, S.Y.; Ni, M.W.; Guo, S.Y. Revealing the common mechanism scutellarin in angina pectoris and ischemic stroke treatment via a network pharmacology approach. *Chin. J. Integr. Med.* **2021**, 62–69. [CrossRef] [PubMed]
105. Gokuladhas, S.; Schierding, W.; Cameron-Smith, D.; Wake, M.; Scotter, E.L.; O'Sullivan, J. Shared regulatory pathways reveal novel genetic correlations between grip strength and neuromuscular disorders. *Front. Genet.* **2020**, *11*, 393. [CrossRef] [PubMed]
106. Costa Sa, A.C.; Madsen, H.; Brown, J.R. Shared molecular signatures across neurodegenerative diseases and herpes viral infections highlights potential mechanisms for maladaptive innate immune responses. *Sci. Rep.* **2019**, *9*, 8795. [CrossRef] [PubMed]
107. Luan, M.; Shang, Z.; Teng, Y.; Chen, X.; Zhang, M.; Lv, H.; Zhang, R. The shared and specific mechanism of four autoimmune diseases. *Oncotarget* **2017**, *8*, 108355–108374. [CrossRef] [PubMed]

108. Landolt, L.; Spagnoli, G.C.; Hertig, A.; Brocheriou, I.; Marti, H.-P. Fibrosis and cancer: Shared features and mechanisms suggest common targeted therapeutic approaches. *Nephrol. Dial. Transplant.* **2020**, in press. [CrossRef] [PubMed]
109. Ormstad, H.; Simonsen, C.S.; Broch, L.; Maes, D.M.; Anderson, G.; Celius, E.G. Chronic fatigue and depression due to multiple sclerosis: Immune-inflammatory pathways, tryptophan catabolites and the gut-brain axis as possible shared pathways. *Mult. Scler. Relat. Disord.* **2020**, *46*, 102533. [CrossRef] [PubMed]
110. Tap, L.; Kirkham, F.A.; Mattace-Raso, F.; Joly, L.; Rajkumar, C.; Benetos, A. Unraveling the links underlying arterial stiffness, bone demineralization, and muscle loss. *Hypertension* **2020**, *76*, 629–639. [CrossRef]
111. Yaron, A.; Schuldiner, O. Common and divergent mechanisms in developmental neuronal remodeling and dying back neurodegeneration. *Curr. Biol.* **2016**, *26*, R628–R639. [CrossRef]
112. Inzelberg, R.; Flash, S.; Friedman, E.; Azizi, E. Cutaneous malignant melanoma and Parkinson disease: Common pathways? *Ann. Neurol.* **2016**, *80*, 811–820. [CrossRef]
113. Zhu, Y.; Ding, X.; She, Z.; Bai, X.; Nie, Z.; Wang, F.; Wang, F.; Geng, X. Exploring shared pathogenesis of Alzheimer's disease and type 2 diabetes mellitus via co-expression networks analysis. *Curr. Alzheimer Res.* **2020**, *17*, 566–575. [CrossRef] [PubMed]
114. Karki, R.; Kodamullil, A.T.; Hofmann-Apitius, M. Comorbidity analysis between Alzheimer's disease and type 2 diabetes mellitus (T2DM) based on shared pathways and the role of T2DM drugs. *J. Alzheimer's Dis.* **2017**, *60*, 721–731. [CrossRef] [PubMed]

Article

Impact of Water Temperature on Heart Rate Variability during Bathing

Jianbo Xu and Wenxi Chen *

Biomedical Information Engineering Laboratory, The University of Aizu, Aizu-Wakamatsu 965-8580, Japan; d8211103@u-aizu.ac.jp
* Correspondence: wenxi@u-aizu.ac.jp

Abstract: Background: Heart rate variability (HRV) is affected by many factors. This paper aims to explore the impact of water temperature (WT) on HRV during bathing. Methods: The bathtub WT was preset at three conditions: i.e., low WT (36–38 °C), medium WT (38–40 °C), and high WT (40–42 °C), respectively. Ten subjects participated in the data collection. Each subject collected five electrocardiogram (ECG) recordings at each preset bathtub WT condition. Each recording was 18 min long with a sampling rate of 200 Hz. In total, 150 ECG recordings and 150 WT recordings were collected. Twenty HRV features were calculated using 1-min ECG segments each time. The k-means clustering analysis method was used to analyze the rough trends based on the preset WT. Analyses of the significant differences were performed using the multivariate analysis of variance of t-tests, and the mean and standard deviation (SD) of each HRV feature based on the WT were calculated. Results: The statistics show that with increasing WT, 11 HRV features are significantly ($p < 0.05$) and monotonously reduced, four HRV features are significantly ($p < 0.05$) and monotonously rising, two HRV features are rising first and then reduced, two HRV features (fuzzy and approximate entropy) are almost unchanged, and vLF power is rising. Conclusion: The WT has an important impact on HRV during bathing. The findings in the present work reveal an important physiological factor that affects the dynamic changes of HRV and contribute to better quantitative analyses of HRV in future research works.

Keywords: water temperature; bathing; ECG; heart rate variability; quantitative analysis; t-test

Citation: Xu, J.; Chen, W. Impact of Water Temperature on Heart Rate Variability during Bathing. *Life* **2021**, *11*, 378. https://doi.org/10.3390/life11050378

Academic Editors: Md. Altaf-Ul-Amin, Shigehiko Kanaya, Naoaki Ono and Ming Huang

Received: 30 March 2021
Accepted: 20 April 2021
Published: 22 April 2021

Publisher's Note: MDPI stays neutral with regard to jurisdictional claims in published maps and institutional affiliations.

1. Introduction

Heart rate variability (HRV) is an important indicator of physical and mental health. The instantaneous HRV rhythm represents a dynamic balance between the sympathetic nervous system (SNS) and parasympathetic nervous system (PNS) branches of the autonomic nervous system (ANS) [1]. Therefore, the quantitative analysis of HRV is considered an effective method for the diagnosis of many cardiac diseases in clinical applications. However, many internal and external factors affect HRV. The internal factors include mental stress, gender, age, and disease, while the external factors include sleep, drugs, alcohol, smoking, and diet.

1.1. HRV and Stress

The SNS branch of the ANS was more activated during states of mental stress [2]; therefore, some literature evaluated mental stress using HRV analyses based on different stressors [3–12]. Some papers confirmed that the HR was significantly increasing during stress states [5,7,13–20], while some papers found that the mean R-R intervals (RRIs) [5,7–10,12,14–16,21] and the square root of the mean of the squares of the successive differences (RMSSD) between adjacent normal to normal intervals (NNs) [8,10,22–29] were significantly reduced during stress states. Kofman et al. discovered that the percentage of low frequency power in total power, pLF, was significantly higher while the percentage of high frequency power in total power, pHF, was significantly reduced during an examination stress state [4]. Melillo et al. found that the LF/HF ratio was significantly higher in the normal estimated glomerular filtration rate [30], while Hjortskov et al. proved that the LF/HF ratio was significantly higher during computer work stress states [3].

1.2. HRV and Gender and Age

The HRV dynamically changes with aging and gender. Ramaekers et al. and Schwartz et al. discovered that some HRV parameters decreased with aging, while Ramaekers et al. confirmed that the gender differences in the HRV parameters only exist in subjects younger than 40 years old [31,32]. Lochner et al. found that women's HRV was significantly lower than men's HRV [33]. Davy et al. observed that physically active women had higher levels of cardiac baroreflex sensitivity and HRV compared with sedentary women regardless of age [34]. Nagy et al. proved that gender differences determined HRV differences from birth, while boys' HR baseline was significantly lower than that for girls [35]. Bonnemeier et al. noted that gender differences in HRV were significantly reduced with aging [36]. Yamasaki et al. discovered that LF was highly determined by aging and the pLF of men was significantly higher than that of women [37].

1.3. HRV and Disease

The HRV differs between healthy people and patients. Wilkowska et al. found that the HRV of depressed patients was significantly lower than that of nondepressed patients [38]. Lutfi and Sukkar showed that lower HRV was associated with higher BP values, putting subjects with such trends at a higher risk of developing hypertension [39]. T. Tombul et al. confirmed that lower HRV in multiple sclerosis patients than that in healthy [40]. D. Gurses et al. observed that some time domain parameters (mean RRI, SDNN, RMSSD, and PNN50) were significantly lower in the thalassaemic patients then that of the healthy subjects [41]. M. Lan et al. found that the mean RRIs significantly increased while LF% and LF/HF significantly decreased in the patients with allergic rhinitis in the sitting position [42]. DelRosso et al. investigated obstructive sleep apnea and found that resulted in increased sympathetic activation during sleep in children [43].

1.4. HRV and Sleep

The sleep has an important impact on the HRV. Herzig et al. discovered that the HF was higher during REM sleep than during slow wave sleep (deep sleep) [44]. Padole and Ingale found that the HRV was distinguishable among the normal, sleeping, and meditation states [45]. Arslan et al. confirmed that the sleep deprivation resulted in a significant decreased in HF, TP, standard deviation (SD) of NN intervals (SDNN), and pNN with concomitant increased in the LF/HF ratio [46]. ÁR. Sűdy et al. confirmed that the HRV during sleep on workdays and free days was significantly different in young healthy men with social jetlag [47].

1.5. HRV and Other Factors

Many other factors also affect HRV. Hynynen et al. proved that the HRV of healthy men was significantly decreased, and the HR was significantly increased at night after marathon or moderate exercise sessions [48]. James et al. learned that the level of HF significantly changed after severe intensity exercise [49]. Zuanetti et al. discovered that HRV significantly varied after patients took antiarrhythmic drugs [50]. Murgia et al. confirmed that HRV significantly increased after smoking cessation [51]. Young et al. learned that diet played an important role in the link between mood and HRV [52]. Latha et al. learned that classical music had a beneficial effect on HRV and reduced medical students' stress levels [53]. Sollers et al. investigated the varying ambient temperature and found it had an important impact on the HRV [54]. Shin proved that ambient temperature induced a significant difference in pulse rate variability compared to HRV, and that difference became greater at a higher ambient temperature [55].

Some previous studies explored the impact of water temperature (WT) on HRV. Mourot et al. and HC. Choo et al. found that immersion in different WT had an important impact on the HRV [56,57]. Y. Kataoka et al. studied the impact of WT on HRV during bathing, but only 38 °C and 41 °C were included, and a few HRV measures were evaluated [58]. F Edelhäuser et al. investigated the effects of whole-body immersion on HRV at three different WTs (33 °C, 36 °C, and 39 °C) [59].

The main purpose of this paper aims to explore the impact of different WTs on HRV during bathing. The experiment was carried out based on the most commonly used WTs in the daily family life, twenty HRV features (included time domain, frequency domain, and non-linear domain) were calculate.

2. Method

2.1. ECG Collection System

The electrocardiogram (ECG) collection system in this study includes four rectangular stainless steel noncontact electrodes, all of them placed on the bathtub wall. When the subject is in the bathtub during bathing, the four noncontact electrodes are near the right foot, right arm, left foot, and left arm, respectively.

The electricity on the skin surface, which is produced by the electrical activity of the heart, arrives in the four noncontact electrodes through the water and three-lead ECG are recorded. The lead I ECG is the potential difference between the left arm (positive) and right arm (negative), the lead II ECG is the potential difference between the left foot (positive) and right arm (negative), and the lead III ECG is the potential difference between the left foot (positive) and left arm (negative). Four shielded wires are, respectively, welded onto the four noncontact electrodes. The three-lead ECG arrives in the ECG collection monitor (Open Brain Computer Interface Biosensing Ganglion Board-OpenBCI Ganglion; OpenBCI, USA) through the shielded wires, and the ECG monitor and the laptop (a MacBook Pro) are connected using a standard Bluetooth 4.n, and all the collected ECG recordings are stored on the laptop. The designed ECG collection system in this study is shown in Figure 1 [60].

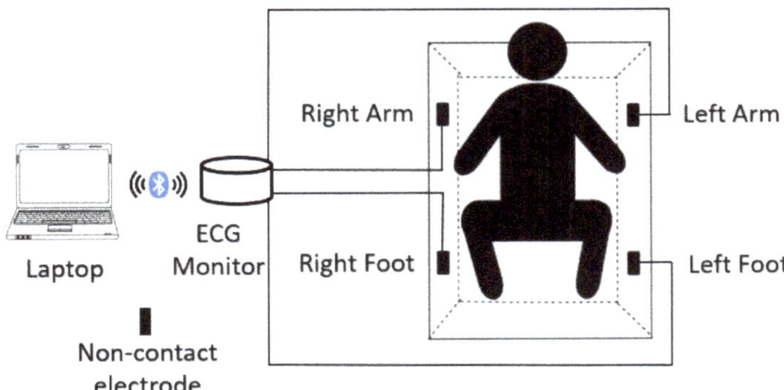

Figure 1. ECG collection system.

2.2. Subjects and ECG Recordings

The ECG recording procedures were approved by the Public University Corporation, the University of Aizu Research Ethics Committee. Written informed consent was obtained from each participant before the experiment.

Ten subjects (seven men and three women) aged 23 to 40 years old (mean ± SD: 28.5 ± 4.78 years) who were students from the University of Aizu participated in the data collection. The BP, body temperature, and body weight were recorded before and after the ECG collection, and the temperature profile for WT change and room temperature were recorded every second during the ECG collection using a temperature monitor (TR-71wb/nw; T&D Corporation, 817-1, Shimadachi, Matsumoto, Nagano, Japan, 390-0852).

The preset bathtub WT includes three conditions: low WT (36–38 °C), medium WT (38–40 °C), and high WT (40–42 °C), respectively. Each subject collected 5 ECG recordings at each preset bathtub WT condition and each recording was 18 min long with a sampling rate of 200 Hz. In total, 150 ECG recordings and 150 temperature recordings were collected during bathing.

2.3. ECG processing

The flowchart for the ECG processing, HRV feature calculation, and statistical analysis is shown in Figure 2.

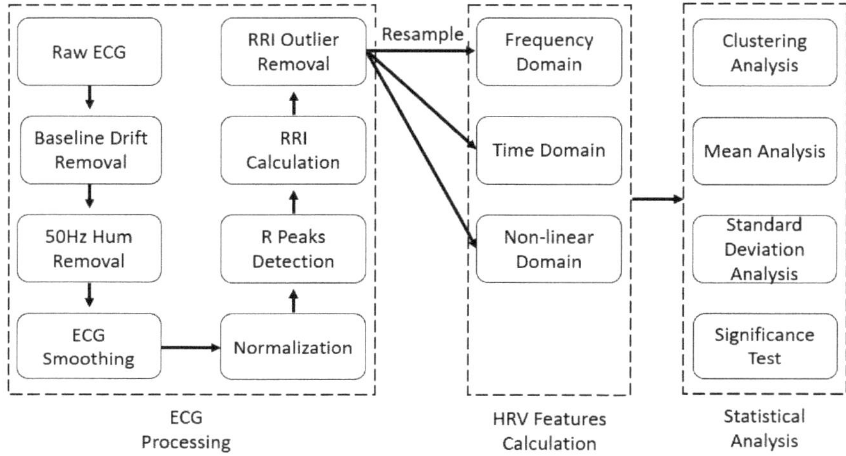

Figure 2. Flowchart for the ECG processing, HRV feature calculation, and statistical analysis.

All data processing and analyses were performed using MATLAB (R2019a). Baseline wandering is obvious in the raw ECG signal due to motion artifacts and respiration from the subjects; therefore, the wandering baseline was removed using the one-dimensional (1-D) wavelet decomposition and reconstruction methods. The Daubechies wavelet at level 10 was used to decompose the raw ECG signal and the baseline wandering approximation coefficient was subtracted from the raw ECG signal after reconstructing at level 8.

Obvious hum noise was also observed in the raw ECG signal; therefore, we performed a spectrum analysis on the raw ECG signal. The spectrum analysis results show that the main frequency component of the hum noise was 50 Hz, which is mainly produced by the electromagnetic interference between the power supply network and equipment [6]. A second-order infinite impulse response digital notch filter was used to remove the 50 Hz hum noise. The numerator and denominator coefficients of the digital notch filter with the notch located at ω and the bandwidth at 0.0071 at the -3 dB level were calculated and the ω must meet the condition of $0.0 < \omega < 1.0$. The difference equation of the digital notch filter is shown in Equation (1).

$$y[n] = \sum_{i=0}^{N} b_i x[n-i] - \sum_{i=1}^{M} a_i y[n-i], \qquad n \geq 0 \tag{1}$$

where $x[n]$ is the filter input, $y[n]$ is the filter output, and a_i and b_i are the numerator and denominator coefficients, respectively, of the digital notch filter.

Next, the 5-point moving average method was used to smoothen the ECG signal. The mathematical formula for the moving average is shown in Equation (2):

$$y[n] = \frac{1}{M} \sum_{j=0}^{M-1} x[n-j]$$ (2)

where $x[n]$ is the input signal, $y[n]$ is the output signal, and M is 5.

Then the ECG was normalized into the range of 0 to 1 using the "mapminmax" function, the R peaks were detected using the "findpeaks" function, the RRI were calculated, and the RRI outliers removed using the 1D 11th order median filter because of its outstanding capability in suppressing the isolated outlier noise without blurring sharp changes in the original signal.

The mathematical formula of the 1D 11th order median filter is shown in Equation (3):

$$y[i] = median\{x[i], i \in w\}$$ (3)

where $x[i]$ is the input signal, $y[i]$ is the output signal, and w is the moving window length. The results for each ECG processing step are shown in Figure 3.

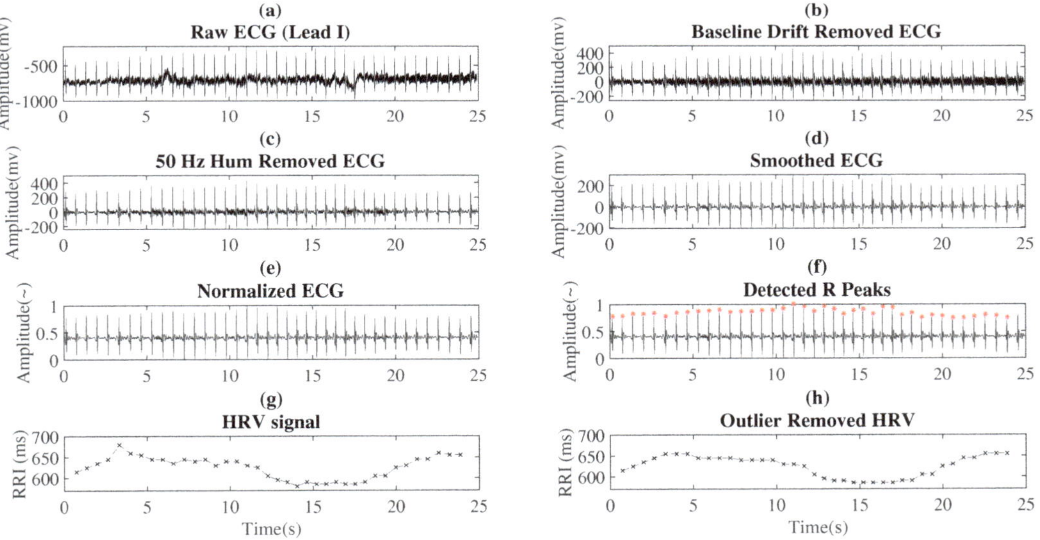

Figure 3. Output of the ECG processing steps.

2.4. HRV Analysis

HRV analysis methods include linear and nonlinear domain analysis methods, where the linear domain includes time and frequency domain methods. The HRV features in the time domain include HR, mean RRI, SDNN, RMSSD between adjacent NNs, SD of the successive differences between adjacent NNs (SDSD), and area under RRI (AURRI). The HRV features in the frequency domain include very LF (VLF) power (0.003–0.04 Hz), LF power (0.04–0.15 Hz), HF power (0.15–0.4 Hz), total power (0–0.4 Hz), pLF, pHF, and the LF/HF ratio. The HRV features in the nonlinear domain include the correlation dimension (D_2), the SD of the Poincare plot perpendicular to the line of identity (SD1), the SD of the Poincare plot along to the line of identity (SD2), the SD1/SD2 ratio, and the sample (SE), fuzzy (FE), and approximate entropies (AE).

Before the frequency features are calculated, a RRI resample is necessary. According to Nyquist's sampling theorem, the sample rate must be more than two times the highest frequency in the original signal. The highest frequency of the HRV is 0.4 Hz; therefore the new resampling rate for RRI was set at 2 Hz. Then, a discrete Fourier transform (DFT was used to calculate the power spectral density (PSD) of the resampled RRI for a N point sequence. Its DFT is shown in Equation (4):

$$X[k] = \sum_{n=0}^{N-1} x[n]e^{-i\frac{2\pi}{N}nk}. \quad (4$$

where $k = 0, 1, 2, \ldots, N-1$, and $i^2 = -1$.

2.5. Statistical Analysis of the HRV Features

Each ECG recording was 18 min in length and segmented into 18 equal parts. 1-min ECG was used to calculate the HRV features each time. Based on the bathtub W the mean and the SD of each HRV feature were calculated, and the *t*-test was used to te for significance. The summary statistic results of each HRV feature are visualized in th clustering results and box plot.

3. Results

The variations of the HRV features based on different WTs are shown in Figure 4. Th smaller dots with blue, yellow, and green colors represent the HRV features calculate based on each preset WT, while the bigger black dots are the average values of the HR features based on each preset WT calculated by the K-means clustering analysis metho For the areas of the dots of HR, the blue area is smallest, the yellow area is biggest, and th green area is medium sized. The low WT condition is very close to the average temperatu (about 36.5 °C) of the normal human body, therefore, the WT stimulation was not stro to the subjects, with a small variation in HR and the SD of HR was 3.38. The higher W has a stronger stimulation to the subjects during bathing. Although the instantaneous H was very fast at this WT condition, the HRV was not the biggest with a SD of HR 4.1 The stimulation for the medium WT condition was bigger than the stimulation for the lo WT condition, but was smaller than the high WT condition. The HRV is obvious with a S of HR 4.65; therefore, the area of the yellow dots was the biggest.

Figure 4 shows that the controlled condition of WT was not serious or uniform. fact, the low WT was not strictly and evenly distributed in the range of (36–38 °C) a was far below the ambient temperature during bathing. The WT of 42 °C was far beyo the ambient temperature during bathing; thus, the WT decreased quickly during the da collection and the WT data at about 42 °C were not as concentrated. It is obvious that t D_2, HF power, total power, pHF, mean RRI, SDNN, RMSSD, SDSD, AURRI, SD1, and S were monotonously reduced with increasing WT, and the pLF, LF/HF, HR, and SD1/S were monotonously rising with increasing WT. A significant difference ($p < 0.05$) was fou among the above HRV features.

The results of significant difference analyses for the 20 HRV features in three analy domains under three WT conditions are visualized in Figure 5. There are some outliers each HRV feature. For the HR, the higher of the WT, the more outliers because the subje experienced stronger stimulation from the higher WT and it is more difficult to adapt WT environment during bathing. The changes in the mean of VLF, LF, SE, FE, and AE w not obvious, and significant differences were not found in these five HRV features.

The details of the statistic results of the 20 HRV features in the time, frequency, and n linear domains are shown in Table 1, where the mean values and SD are listed, and pairwise statistically significant differences between each WT condition are calcula The significant difference analysis was performed via the multivariate analysis using *t*-test variance method, where *p1* represents the significant difference between low a medium WT conditions, *p2* represents the significant difference between medium and h WT conditions, and *p3* represents the significant difference between low and high conditions. With the increasing WT, the SD of the HR, mean RRI, AURRI, pLF, pHF, LF/

ratio, and SD1/SD2 are first rising and then reduced, and the SD of LF, HF, TP are first reduced and then rising.

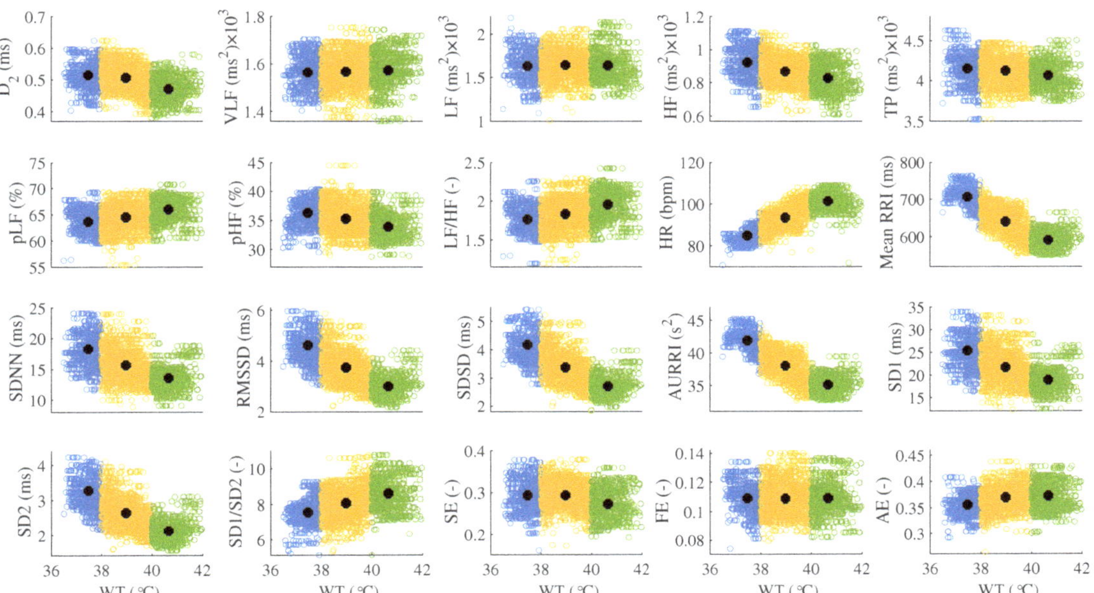

Figure 4. Feature trends for 20 HRV features in three analysis domains under three WT conditions.

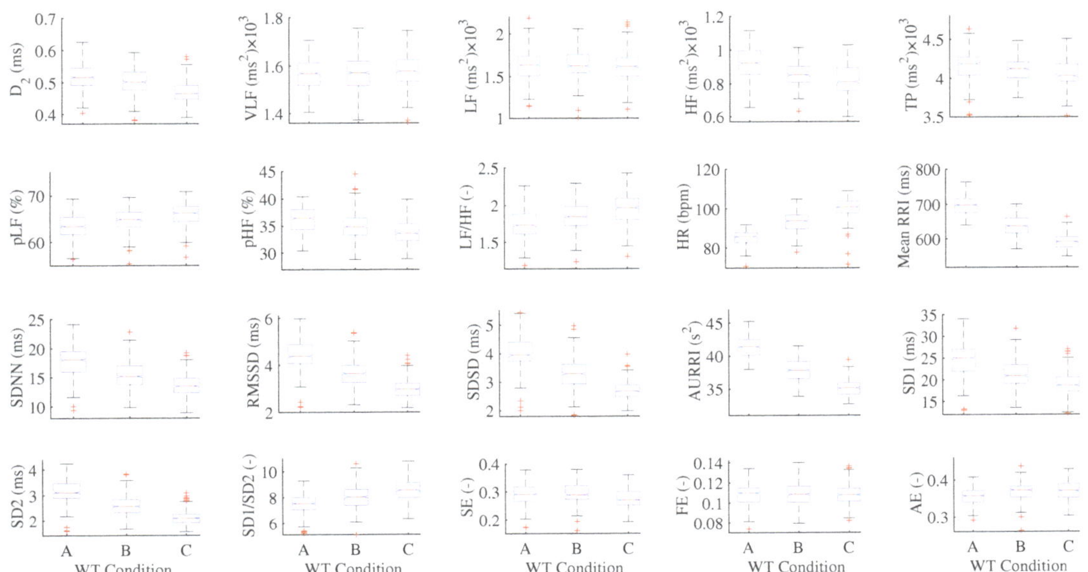

Figure 5. Analysis of significant differences for 20 HRV features in three analysis domains under three WT conditions: A = (36–38) °C, B = (38–40) °C, and C = (40–42) °C.

Table 1. The statistic results of the HRV features based on different bathtub WT.

HRV Features		Features	(36 38] °C		(38 40] °C		(40 42] °C		*p* Value		
		Trend	Mean	SD	Mean	SD	Mean	SD	*p1*	*p2*	*p3*
Time	HR (bpm)	↑↑	85.55	3.38	94.07	4.65	101.30	4.17	0	0	0
Domain	Mean RRI (ms)	↓↓	699.93	27.53	637.04	30.67	591.15	22.42	0	0	0
	SDNN (ms)	↓↓	17.99	2.94	15.48	2.48	13.70	1.87	0	0	0
	RMSSD (ms)	↓↓	4.54	0.74	3.69	0.57	3.03	0.42	0	0	0
	SDSD (ms)	↓↓	4.11	0.65	3.32	0.50	2.72	0.35	0	0	0
	AURRI (s^2)	↓↓	41.51	1.62	37.82	1.80	35.12	1.32	0	0	0
Frequency	VLF Power (ms^2)	↑	1561.77	78.82	1567.90	87.13	1573.03	80.22	0.07	0.14	0
Domain	LF Power (ms^2)	∧	1631.05	182.50	1646.35	159.13	1633.92	175.79	0.04	0.11	0.74
	HF Power (ms^2)	↓↓	922.03	97.45	859.79	68.12	826.23	86.79	0	0	0
	Total Power (ms^2)	↓↓	4157.74	199.76	4114.04	158.49	4065.92	159.75	0	0	0
	pLF (%)	↑↑	63.53	2.37	64.75	2.58	65.91	2.29	0	0	0
	pHF (%)	↓↓	36.36	2.37	35.13	2.45	33.97	2.20	0	0	0
	LF/HF (-)	↑↑	1.75	0.18	1.85	0.20	1.95	0.19	0	0	0
Non-linear	D_2 (ms)	↓↓	0.52	0.04	0.51	0.04	0.47	0.04	0	0	0
Domain	SD1 (ms)	↓↓	24.94	4.10	21.51	3.48	19.11	2.62	0	0	0
	SD2 (ms)	↓↓	3.21	0.53	2.61	0.40	2.14	0.30	0	0	0
	SD1/SD2 (-)	↑↑	7.63	0.82	8.03	0.86	8.65	0.85	0	0	0
	SE (-)	∧	0.29	0.04	0.30	0.03	0.27	0.03	0.15	0	0
	FE (-)	∼	0.11	0.01	0.11	0.01	0.11	0.01	0.21	0.04	0.43
	AE (-)	∼	0.36	0.02	0.37	0.02	0.37	0.02	0	0.14	0

↓↓ (↑↑): Significantly reduced (increase) with increasing WT $p < 0.05$; ↓ (↑): Reduced (increase) with increasing WT $p > 0.05$; ∧: Increased first and then reduced with increasing WT; ∼: Unobvious change; $p1$= between (36–38) °C and (38–40) °C; $p2$= between (38–40) °C and (40–42)°C; $p3$= between (36–38) °C and (40–42) °C. (36–38)°C: $36 < WT \leq 38$ °C (38–40) °C: $38 < WT \leq 40$ °C; (40–42) °C: $40 < WT \leq 42$ °C;

4. Discussion

As a noninvasive, rapid, and accurate tool in the evaluation of the cardiac autonomic balance modulation activity, heart rate variety (HRV) has been a hot research topic in recent years. This study explored the impact of different water temperature (WT) on HRV during bathing. With the rises of WT, the HR in medium and high WT increased by 6.53% and 15.78%, respectively, compared with the low WT, which reflects a decreased vagal modulation. The significantly and monotonously reduced SDNN with increasing WT shows a significantly reduced whole HRV fluctuation, which is highly consistent with the significantly and monotonously reduced total spectral power (0–0.4 Hz). The LF power (0.04–0.15 Hz) in the PSD reflects both SNS and PNS activities, while the HF power (0.15–0.4 Hz) in the PSD reflects the PNS activity, and the LF/HF ratio represents the balance between the SNS and PNS activities. With the increasing WT, the LF and HF are significantly and monotonously reduced, which reflects that SNS and PNS activities are enhanced significantly. The increased LF/HF ratio shows that the ratio of the cardiac sympathetic to parasympathetic tones (i.e., the sympathovagal balance) was enhanced significantly, which shows that the stimulation of high WT on the subject was also enhanced significantly. The stimulation on the subject under high WT increased by 6.43 and 5.20% over the low and medium WT, as shown in Table 1. Furthermore, the HRV feature of AURRI was newly defined in this paper and its unit is s^2. The AURRI reflects the fluctuation of HRV signal over time: i.e., with the increasing WT, the mean RRI and AURRI are reduced. In lower WT condition, the parasympathetic activity is dominant. With the WT increasing, our findings show decreased HRV complexity, which induce obvious shift of ANS balance towards the sympathetic activation associated with vagal withdrawal. Therefore, the higher WT can induce a stronger response of physiological allostatic regulatory, which is often accompanied by an enhanced cardiac sympathoexcitation associated with a vagal withdrawal. From the healthcare perspective, to reduce the sudden onset possibility of cardiac and cardiovascular complications or diseases during bathing, it is more dangerous to choose a higher WT condition for the patients.

The same WTs which belonging to different WT change processes induce different impacts on HRV. For example, if the WT drops from 40 °C to 38 °C during the data collection process, the subject will feel very uncomfortable in the first minute and need a longer time to adapt to the WT environment. However, if the WT increases from 38 °C to 40 °C during the data collection process, the subject will adapt to the WT environment easily. Even if the WT reaches 40 °C, the subjects will not feel very uncomfortable because they have adapted to this WT environment. The WT of 40 °C appeared during two different processes, but had very different instantaneous effects on the HRV and their physiological meanings were also different in these processes. Therefore, some outliers appear in the box plot as shown in Figure 5.

According to experiment records, the difference from other subjects was that Subjects 4, 7, and 8 did not feel very uncomfortable even in the higher WT (40–42 °C).The slopes of the variety of HR are smaller than that of the other seven subjects, as shown in Table 2. Specifically, Subject 7 preferred the higher WT and the change in their HR was not as obvious as in the other subjects, as shown in Figure 6. The questionnaire showed that Subject 7 often participated regular physical activities. Regular exercise could make the sympathetic-adrenal system more easily excited, thus enhancing cardiovascular, hormonal and metabolic responses, further affecting body temperature regulation, water-electrolyte interface, muscle contraction performance, etc., thus ensuring blood perfusion, oxygen, and nutrient supply and elimination of metabolites in organs and tissues throughout the body. There was evidence that exercise could reduce the sympatho-excitation and sympathetic outflow, and the baroreflex-mediated was also suppressed. Therefore, compared with other subjects, Subject 7 demonstrated higher HRV, and their reaction to higher WT indicated a great adaptability of the ANS.

Table 2. The slope of the variety of HR with increasing WT.

Subject	1	2	3	4	5	6	7	8	9	10
Slope	10.71	4.59	6.72	4.34	6.36	7.48	-0.41	3.46	8.91	6.83

Subjects 1, 6, and 9 were very sensitive to changes in WT and especially could not tolerate the high bathtub WT (40–42 °C). They felt more comfortable during 4th–11th min on the data collection process. The first 3 min are the adaptation phase. During this stage, their foreheads quickly began to sweat a lot. For the other seven subjects, their adaptation phases were the first 1 min and they began to sweat more after the first 10 min in the same high WT environment. The body weights of these three subjects were reduced more after data collection than in the other seven subjects. From this finding, we speculate that people who are more sensitive to temperature changes are less able to withstand water and WT pressures, and they are more likely to suffer from higher mental and physical stress during the higher WT condition.

Subjects 2, 3, 5, and 10 felt a little uncomfortable, but could tolerate the high WT (40–42 °C). All ten subjects could quickly adapt to the low WT (36–38 °C) and they felt more relaxed and comfortable in the medium WT (38–40 °C). Except for three subjects who were very sensitive to the WT changes, the other seven subjects did not feel uncomfortable.

Although some discoveries were revealed in this paper, there are also some limitations. First, the set-up of the experiment is stressful itself, and therefore may create an additional bias. Although informed consent was obtained and the data collection process was told in detail to each subject, as well as let each subject had a five-minute rest before the data collection, some subjects were still a little nervous at the beginning of data collection. Furthermore, the sensitivity to external stimuli of each subject was different, the water pressure on the chest and thermal stimulus on hemodynamics also induced different stress to each subject, which would induce some additional biases to the results. Therefore, the mental stress factor should be also taken into consideration at the same time to evaluate the impact of WT on HRV. Second, the number of subjects is too small and the subjects should include older people and children, in addition to healthy and unhealthy individuals and different ethnicities. Third, the change ranges for WT during the data collection were too big. Due to poor insulation measures, the WT was relatively divergent during the data collection process. Thus, the HRV analysis should be performed based on several smaller ranges of WT. Fourth, the data collection environment was inconsistent for all the subjects. For example, when the WT is between 40 °C to 42 °C, some subjects could endure the high WT environment, while other subjects felt too uncomfortable to endure the high WT environment. Therefore, to be safe, we must give these subjects a fan to blow a refreshing cool breeze on themselves in this situation. Fifth, during the data-processing stage, the median filter was used three times to remove the outliers of the RRI signal. The skin surface electricity is very weak, in the millivolts. The gentle movement of the four limbs will induce relatively large fluctuations in the ECG amplitude. Therefore, the raw ECG signal includes some noise and there are some outliers in the R peaks detection and RRI signal results. If the median filter is only used once to remove the RRI outliers, then either only the outliers with big amplitude can be removed or there is a gross distortion in the RRI signal after the outliers are removed. Therefore, the median filter was used to remove the outliers with big, median, and small amplitudes, respectively.

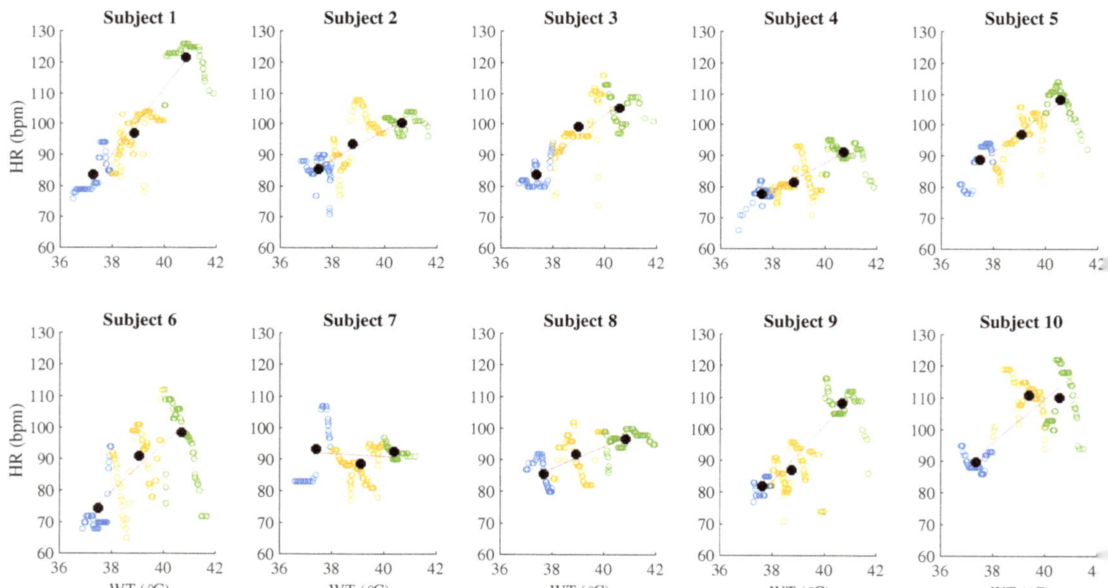

Figure 6. The variety of HR and its fitted curve at one degree for each subject under three WT conditions.

5. Conclusions and Future Work

This paper explores the impact of different WTs on HRV during bathing. With the WTs increasing, some HRV features are significantly and monotonously reduced or rising, which induces the change of dynamic balance between the SNS and PNS branches of the ANS. The findings in the present study provide important reference significance in many practical aspects which need to evaluate the amount of disturbance of homeostasis induced by WTs. For example, we can affect the HRV by changing the WTs to set an optimal environment during bathing. Only when the SNS and PNS activities are controlled a certain range can the people feel relaxed and comfortable.

In future research works, we will further explore the HRV levels of healthy subjects and patients, especially the patients with cardiac diseases (such as arrhythmia, myocardial ischemia, and coronary heart disease), and then design an automated and accurate WTs control system to affect the HRV by changing the WTs so that the HRV level is indirectly controlled in a safe and comfortable range based on individual health condition, which would appropriately reduce the possibility of sudden onset of cardiac disease during bathing. Moreover, in order to achieve the purpose of lifelong healthcare, we will also explore how to apply the cutting-edge blockchain technology in the long-term tracking of ECG data during bathing for the big data collection and analysis [62,63]. Another particularly crucial research topic is the physiological signal encryption and secure transmission related to the privacy protection, some emerging technologies provide a valuable reference [64,65].

Author Contributions: Conceived and designed the idea: J.X., W.C.; data collection and analysis, paper writing, and bibliographic research: J.X.; supervision and manuscript revision: W.C. All authors have read and agreed to the published version of the manuscript

Funding: This study was supported in part by the University of Aizu's Competitive Research Fund (2020-P-24).

Institutional Review Board Statement: The ECG recording procedures were approved by the Public University Corporation, the University of Aizu Research Ethics Committee.

Informed Consent Statement: Written informed consent was obtained from each participant before the experiment.

Data Availability Statement: Data are available from the biomedical information engineering laboratory (contact via e-mail: wenxi@u-aizu.ac.jp) for researchers who meet the criteria for access to confidential data.

Acknowledgments: The authors thank all participants for their cooperation during the data collection.

Conflicts of Interest: The authors declare no conflict of interest. The funders had no role in the design of the study; in the collection, analyses, or interpretation of data; in the writing of the manuscript, or in the decision to publish the results.

References

Saul, J.P. Beat-to-beat variations of heart rate reflect modulation of cardiac autonomic outflow. *Physiology* **1990**, *5*, 32–37. [CrossRef]

Van Praag, H. Crossroads of corticotropin releasing hormone, corticosteroids and monoamines. *Neurotox. Res.* **2002**, *4*, 531–556. [CrossRef]

Hjortskov, N.; Rissén, D.; Blangsted, A.K.; Fallentin, N.; Lundberg, U.; Søgaard, K. The effect of mental stress on heart rate variability and blood pressure during computer work. *Eur. J. Appl. Physiol.* **2004**, *92*, 84–89. [CrossRef]

Kofman, O.; Meiran, N.; Greenberg, E.; Balas, M.; Cohen, H. Enhanced performance on executive functions associated with examination stress: Evidence from task-switching and Stroop paradigms. *Cogn. Emot.* **2006**, *20*, 577–595. [CrossRef]

Vuksanović, V.; Gal, V. Heart rate variability in mental stress aloud. *Med. Eng. Phys.* **2007**, *29*, 344–349. [CrossRef]

Li, Z.; Snieder, H.; Su, S.; Ding, X.; Thayer, J.F.; Treiber, F.A.; Wang, X. A longitudinal study in youth of heart rate variability at rest and in response to stress. *Int. J. Psychophysiol.* **2009**, *73*, 212–217. [CrossRef]

Schubert, C.; Lambertz, M.; Nelesen, R.; Bardwell, W.; Choi, J.B.; Dimsdale, J. Effects of stress on heart rate complexity—A comparison between short-term and chronic stress. *Biol. Psychol.* **2009**, *80*, 325–332. [CrossRef]

Tharion, E.; Parthasarathy, S.; Neelakantan, N. Short-term heart rate variability measures in students during examinations. *Natl. Med. J. India* **2009**, *22*, 63–66.

Lackner, H.K.; Papousek, I.; Batzel, J.J.; Roessler, A.; Scharfetter, H.; Hinghofer-Szalkay, H. Phase synchronization of hemodynamic variables and respiration during mental challenge. *Int. J. Psychophysiol.* **2011**, *79*, 401–409. [CrossRef]

Taelman, J.; Vandeput, S.; Vlemincx, E.; Spaepen, A.; Van Huffel, S. Instantaneous changes in heart rate regulation due to mental load in simulated office work. *Eur. J. Appl. Physiol.* **2011**, *111*, 1497–1505. [CrossRef]

Traina, M.; Cataldo, A.; Galullo, F.; Russo, G. Heart rate variability in healthy subjects. *Minerva Psichiatr.* **2011**, *52*, 227–231.

Visnovcova, Z.; Mestanik, M.; Javorka, M.; Mokra, D.; Gala, M.; Jurko, A.; Calkovska, A.; Tonhajzerova, I. Complexity and time asymmetry of heart rate variability are altered in acute mental stress. *Physiol. Meas.* **2014**, *35*, 1319. [CrossRef]

González-Camarena, R.; Carrasco-Sosa, S.; Román-Ramos, R.; Gaitán-González, M.J.; Medina-Bañuelos, V.; Azpiroz-Leehan, J. Effect of static and dynamic exercise on heart rate and blood pressure variabilities. *Med. Sci. Sport. Exerc.* **2000**, *32*, 1719–1728. [CrossRef] [PubMed]

14. Salahuddin, L.; Cho, J.; Jeong, M.G.; Kim, D. Ultra short term analysis of heart rate variability for monitoring mental stress in mobile settings. In Proceedings of the 2007 29th Annual International Conference of the IEEE Engineering in Medicine and Biology Society, Lyon, France, 22–26 August 2007; pp. 4656–4659.

15. Costin, R.; Rotariu, C.; Pasarica, A. Mental stress detection using heart rate variability and morphologic variability of EeG signals. In Proceedings of the 2012 International Conference and Exposition on Electrical and Power Engineering, New York, NY, USA, 25–27 October 2012; pp. 591–596.

16. Ruediger, H.; Seibt, R.; Scheuch, K.; Krause, M.; Alam, S. Sympathetic and parasympathetic activation in heart rate variability in male hypertensive patients under mental stress. *J. Hum. Hypertens.* **2004**, *18*, 307–315. [CrossRef] [PubMed]

17. Owen, N.; Steptoe, A. Natural killer cell and proinflammatory cytokine responses to mental stress: Associations with heart rate and heart rate variability. *Biol. Psychol.* **2003**, *63*, 101–115. [CrossRef]

18. Custodis, F.; Gertz, K.; Balkaya, M.; Prinz, V.; Mathar, I.; Stamm, C.; Kronenberg, G.; Kazakov, A.; Freichel, M.; Böhm, M.; et al. Heart rate contributes to the vascular effects of chronic mental stress: Effects on endothelial function and ischemic brain injury in mice. *Stroke* **2011**, *42*, 1742–1749. [CrossRef]

19. Jouven, X.; Schwartz, P.J.; Escolano, S.; Straczek, C.; Tafflet, M.; Desnos, M.; Empana, J.P.; Ducimetière, P. Excessive heart rate increase during mild mental stress in preparation for exercise predicts sudden death in the general population. *Eur. Heart J.* **2009**, *30*, 1703–1710. [CrossRef]

20. Yadav, A.; Awasthi, N.; Gaur, K.; Yadav, K. Comparison of Heart Rate Variability during Physical and Mental Stress in Type 'A' and Type 'B' personality: An Interventional Study. *Int. Multispecialty J. Health (IMJH)* **2019**, *5*, 133–140.

21. Papousek, I.; Nauschnegg, K.; Paechter, M.; Lackner, H.K.; Goswami, N.; Schulter, G. Trait and state positive affect and cardiovascular recovery from experimental academic stress. *Biol. Psychol.* **2010**, *83*, 108–115. [CrossRef]

22. Melzig, C.A.; Weike, A.I.; Hamm, A.O.; Thayer, J.F. Individual differences in fear-potentiated startle as a function of resting heart rate variability: Implications for panic disorder. *Int. J. Psychophysiol.* **2009**, *71*, 109–117. [CrossRef]

23. Hauschildt, M.; Peters, M.J.; Moritz, S.; Jelinek, L. Heart rate variability in response to affective scenes in posttraumatic stress disorder. *Biol. Psychol.* **2011**, *88*, 215–222. [CrossRef] [PubMed]

24. Lakusic, N.; Fuckar, K.; Mahovic, D.; Cerovec, D.; Majsec, M.; Stancin, N. Characteristics of heart rate variability in war veteran with post-traumatic stress disorder after myocardial infarction. *Mil. Med.* **2007**, *172*, 1190–1193. [CrossRef] [PubMed]

25. Guédon-Moreau, L.; Ducrocq, F.; Molenda, S.; Duhem, S.; Salleron, J.; Chaudieu, I.; Bert, D.; Libersa, C.; Vaiva, G. Temporal analysis of heart rate variability as a predictor of post traumatic stress disorder in road traffic accidents survivors. *J. Psychiatr. Re* **2012**, *46*, 790–796.

26. Alvares, G.A.; Quintana, D.S.; Kemp, A.H.; Van Zwieten, A.; Balleine, B.W.; Hickie, I.B.; Guastella, A.J. Reduced heart rate variability in social anxiety disorder: associations with gender and symptom severity. *PLoS ONE* **2013**, *8*, e70468. [CrossRef]

27. Martens, E.; Nyklicek, I.; Szabo, B.; Kupper, N. Depression and anxiety as predictors of heart rate variability after myocardial infarction. *Psychol. Med.* **2008**, *38*, 375–383. [CrossRef]

28. Paritala, S.A. *Effects of Physical and Mental Tasks on Heart Rate Variability*; Louisiana State University: Baton Rouge, LA, USA, 2009; Volume 3928, pp. 1–85.

29. Melillo, P.; De Luca, N.; Bracale, M.; Pecchia, L. Classification tree for risk assessment in patients suffering from congestive heart failure via long-term heart rate variability. *IEEE J. Biomed. Health Inform.* **2013**, *17*, 727–733. [CrossRef]

30. Melillo, P.; Izzo, R.; De Luca, N.; Pecchia, L. Heart rate variability and target organ damage in hypertensive patients. *BMC Cardiovasc. Disord.* **2012**, *12*, 1–11. [CrossRef]

31. Ramaekers, D.; Ector, H.; Aubert, A.; Rubens, A.; Van de Werf, F. Heart rate variability and heart rate in healthy volunteers. Is the female autonomic nervous system cardioprotective? *Eur. Heart J.* **1998**, *19*, 1334–1341. [CrossRef]

32. Schwartz, J.B.; Gibb, W.J.; Tran, T. Aging effects on heart rate variation. *J. Gerontol.* **1991**, *46*, M99–M106. [CrossRef]

33. Lochner, A.; Crooks, P.; Gordon Finlay, M.; Erwin, J.A.; Honn, K. *Effect of Sleep Deprivation and Sleep Recovery on Heart Rate and Heart Rate Variability in Males Versus Females*; Eastern Washington University: Cheney, WA, USA, 2020; Volume 40, pp. 1–2.

34. Davy, K.P.; Desouza, C.A.; Jones, P.P.; Seals, D.R. Elevated heart rate variability in physically active young and older adult women. *Clin. Sci.* **1998**, *94*, 579–584. [CrossRef]

35. Nagy, E.; Orvos, H.; Bárdos, G.; Molnár, P. Gender-related heart rate differences in human neonates. *Pediatr. Res.* **2000**, *47*, 778–782. [CrossRef]

36. Bonnemeier, H.; Wiegand, U.K.; Brandes, A.; Kluge, N.; Katus, H.A.; Richardt, G.; Potratz, J. Circadian profile of cardiac autonomic nervous modulation in healthy subjects: Differing effects of aging and gender on heart rate variability. *J. Cardiovasc. Electrophysiol.* **2003**, *14*, 791–799. [CrossRef]

37. Yamasaki, Y.; Kodama, M.; Matsuhisa, M.; Kishimoto, M.; Ozaki, H.; Tani, A.; Ueda, N.; Ishida, Y.; Kamada, T. Diurnal heart rate variability in healthy subjects: Effects of aging and sex difference. *Am. J. Physiol.-Heart Circ. Physiol.* **1996**, *271*, H303–H310. [CrossRef] [PubMed]

38. Wilkowska, A.; Rynkiewicz, A.; Wdowczyk, J.; Landowski, J.; Cubała, W.J. Heart rate variability and incidence of depression during the first six months following first myocardial infarction. *Neuropsychiatr. Dis. Treat.* **2019**, *15*, 1951–1956. [CrossRef]

39. Lutfi, M.F.; Sukkar, M.Y. Effect of blood pressure on heart rate variability. *Khartoum. Med. J.* **2011**, *4*, 548–553.

40. Tombul, T.; Anlar, O.; Tuncer, M.; Huseyinoglu, N.; Eryonucu, B. Impaired heart rate variability as a marker of cardiovascular autonomic dysfunction in multiple sclerosis. *Acta Neurol. Belg.* **2011**, *111*, 116–120.

1. Gurses, D.; Ulger, Z.; Levent, E.; Aydinok, Y.i.; Ozyurek, A.R. Time domain heart rate variability analysis in patients with thalassaemia major. *Acta Cardiol.* **2005**, *60*, 477–481. [CrossRef]
2. Lan, M.Y.; Lee, G.S.; Shiao, A.S.; Ko, J.H.; Shu, C.H. Heart rate variability analysis in patients with allergic rhinitis. *Sci. World J.* **2013**, *2013*, 1–4. [CrossRef]
3. DelRosso, L.M.; Mogavero, M.P.; Ferri, R. Effect of Sleep Disorders on Blood Pressure and Hypertension in Children. *Curr. Hypertens. Rep.* **2020**, *22*, 1–7. [CrossRef]
4. Herzig, D.; Eser, P.; Omlin, X.; Riener, R.; Wilhelm, M.; Achermann, P. Reproducibility of heart rate variability is parameter and sleep stage dependent. *Front. Physiol.* **2018**, *8*, 1–10. [CrossRef]
5. Padole, A.; Ingale, V.V. Investigating Effect of Sleep and Meditation on HRV and Classification using ANN. In Proceedings of the 2019 10th International Conference on Computing, Communication and Networking Technologies (ICCCNT), Kanpur, India, 6–8 July 2019; pp. 1–6.
6. Arslan, M.; Welcome, M.O.; Dane, S. The Effect of Sleep Deprivation on Heart Rate Variability in Shift Nurses. *J. Res. Med. Dent. Sci.* **2019**, *7*, 45–52.
7. Sűdy, Á.R.; Ella, K.; Bódizs, R.; Káldi, K. Association of social jetlag with sleep quality and autonomic cardiac control during sleep in young healthy men. *Front. Neurosci.* **2019**, *13*, 1–10. [CrossRef]
8. Hynynen, E.; Vesterinen, V.; Rusko, H.; Nummela, A. Effects of moderate and heavy endurance exercise on nocturnal HRV. *Int. J. Sports Med.* **2010**, *31*, 428–432. [CrossRef]
9. James, D.V.; Munson, S.C.; Maldonado-Martin, S.; De Ste Croix, M.B. Heart rate variability: Effect of exercise intensity on postexercise response. *Res. Q. Exerc. Sport* **2012**, *83*, 533–539. [CrossRef]
10. Zuanetti, G.; Latini, R.; Neilson, J.M.; Schwartz, P.J.; Ewing, D.J.; Group, T.A.D.E. Heart rate variability in patients with ventricular arrhythmias: effect of antiarrhythmic drugs. *J. Am. Coll. Cardiol.* **1991**, *17*, 604–612. [CrossRef]
11. Murgia, F.; Melotti, R.; Foco, L.; Gögele, M.; Meraviglia, V.; Motta, B.; Steger, A.; Toifl, M.; Sinnecker, D.; Müller, A.; et al. Effects of smoking status, history and intensity on heart rate variability in the general population: The CHRIS study. *PLoS ONE* **2019**, *14*, e0215053. [CrossRef]
12. Young, H.A.; Cousins, A.L.; Watkins, H.T.; Benton, D. Is the link between depressed mood and heart rate variability explained by disinhibited eating and diet? *Biol. Psychol.* **2017**, *123*, 94–102. [CrossRef]
13. Latha, R.; Srikanth, S.; Sairaman, H.; Dity, N.R.E. Effect of music on heart rate variability and stress in medical students. *Int. J. Clin. Exp. Physiol.* **2014**, *1*, 131–134. [CrossRef]
14. Sollers, J.J.; Sanford, T.A.; Nabors-Oberg, R.; Anderson, C.A.; Thayer, J.F. Examining changes in HRV in response to varying ambient temperature. *IEEE Eng. Med. Biol. Mag.* **2002**, *21*, 30–34. [CrossRef] [PubMed]
15. Shin, H. Ambient temperature effect on pulse rate variability as an alternative to heart rate variability in young adult. *J. Clin. Monit. Comput.* **2016**, *30*, 939–948. [CrossRef]
16. Mourot, L.; Bouhaddi, M.; Gandelin, E.; Cappelle, S.; Dumoulin, G.; Wolf, J.P.; Rouillon, J.D.; Regnard, J. Cardiovascular autonomic control during short-term thermoneutral and cool head-out immersion. *Aviat. Space Environ. Med.* **2008**, *79*, 14–20. [CrossRef]
17. Choo, H.C.; Nosaka, K.; J Peiffer, J.; Ihsan, M.; C Yeo, C.; R Abbiss, C. Effect of water immersion temperature on heart rate variability following exercise in the heat. *Kinesiology* **2018**, *50*, 67–74.
18. Kataoka, Y.; Yoshida, F. The change of hemodynamics and heart rate variability on bathing by the gap of water temperature. *Biomed. Pharmacother.* **2005**, *59*, S92–S99. [CrossRef]
19. Edelhäuser, F.; Goebel, S.; Scheffer, C.; Cysarz, D. P02. 181. Heart rate variability and peripheral temperature during whole body immersion at different water temperatures. *BMC Complement. Altern. Med.* **2012**, *12*, 1. [CrossRef]
20. Xu, J.; Cui, P.; Chen, W. ECG-based Identity Validation during Bathing in Different Water Temperature. In Proceedings of the 2020 42nd Annual International Conference of the IEEE Engineering in Medicine & Biology Society (EMBC), Montreal, QC, Canada, 20–24 July 2020; pp. 5276–5279.
21. Gupta, G.; Mehra, R. Design analysis of IIR filter for power line interference reduction in ECG signals. *Int. J. Eng. Res. Appl.* **2013**, *3*, 1309–1316.
22. Wang, W.; Su, C. Ccbrsn: A system with high embedding capacity for covert communication in bitcoin. In *IFIP Advances in Information and Communication Technology, Proceedings of the IFIP International Conference on ICT Systems Security and Privacy Protection, Maribor, Slovenia, 21–23 September 2020*; Springer: Cham, Switzerland, 2020; pp. 324–337.
23. Wang, W.; Huang, H.; Zhang, L.; Su, C. Secure and efficient mutual authentication protocol for smart grid under blockchain. *Peer-to-Peer Netw. Appl.* **2020**, 1–13. [CrossRef]
24. Zhang, L.; Zhang, Z.; Wang, W.; Jin, Z.; Su, Y.; Chen, H. Research on a Covert Communication Model Realized by Using Smart Contracts in Blockchain Environment. *IEEE Syst. J.* **2021**, *15*, 1–12.
25. Zhang, L.; Zou, Y.; Wang, W.; Jin, Z.; Su, Y.; Chen, H. Resource allocation and trust computing for blockchain-enabled edge computing system. *Comput. Secur.* **2021**, *105*, 102249. [CrossRef]

MDPI
St. Alban-Anlage 66
4052 Basel
Switzerland
Tel. +41 61 683 77 34
Fax +41 61 302 89 18
www.mdpi.com

Life Editorial Office
E-mail: life@mdpi.com
www.mdpi.com/journal/life